P9-AOO-888

CLINICAL NEUROSCIENCE

From Neuroanatomy to Psychodynamics

CLINICAL NEUROSCIENCE

From Neuroanatomy to Psychodynamics

Jay E. Harris, M.D.
Assistant Professor of Psychiatry
New York Medical College

Unit Chief
Department of Psychiatry
Metropolitan Hospital, New York

Edited by Jean A. Harris, Ph.D.

HUMAN SCIENCES PRESS, INC.
72 FIFTH AVENUE
NEW YORK, N.Y. 10011

72 Fifth Avenue, New York, New York 10011

Printed in the United States of America
987654321

Library of Congress Cataloging in Publication Data

Harris, Jay,
 Clinical neuroscience

 Includes bibliographies and index.
 1. Neuropsychiatry. 2. Brain. 3. Mental illness.
I. Harris, Jean. II. Title. III. Title: Neuroanatomy to psychodynamics. [DNLM: 1. Brain—physiology.
2. Brain—physiopathology. 3. Mental disorders.
4. Psychoanalytic Theory. WM 100 H314c]
RC343.H26 1985 616.89 84-28848
ISBN 0-89885-238-2

CONTENTS

FOREWORD

The prospect of developing a unified theory of mind and brain has intrigued psychiatrists since the publication of Freud's *Project for a Scientific Psychology*. Dr. Harris, having identified himself with that tradition, has chosen to set himself no small task. He clearly intends to update Freud's "*Project*" in light of recent developments both in psychoanalytic theory and in the neurosciences. In doing so he is prepared to provide the reader with a way of thinking about the interrelations that must exist between an individual's biology and phenomenology.

In order to achieve these ends the author must acquaint us with pertinent findings in such diverse areas as neurochemistry and psychopharmacology, cognition and neuropsychology, neurophysiology and behavioral neurology, and ego psychology and psychopathology. As you might imagine, just keeping all of these balls in the air at one time is no mean feat. But the larger inherent problem in such an undertaking is how does one get these facts and fancies to cohere into a semblance of an intelligible pattern that the reader can readily apprehend.

The history of ideas really offers few solutions to such a problem. The ones most commonly encountered tend to derive from the root metaphor known as Organism and are evident in the thinking of Hughling-Jackson, Kurt Goldstein, and A. Luria in neuropsychology, Freud in the development of psychoanalytic theory and Piaget in cognitive development. Not surprisingly, each of these eminent thinkers was concerned with the relation of mind and brain as it related to their particular disciplines. I think it fair to say that Dr. Harris' own creative synthesis both draws upon and follows in this tradition.

Briefly, Organism and its more recent offspring General Systems Theory, give rise to a meta-system that enables one to describe the operations and interactions of very different phenomena that occur within living systems. Thus, in principle, it makes no difference whether one chooses to discuss neurochemistry, information processing, or the symbol-generating capacity of a person; Organism provides a language system capable of characterizing each of these domains or levels of activity, as well as their interaction.

Organism begins with the assertion that *change* is axiomatic in living systems. Development marks the successful accommodation to change while the inability to cope with change is revealed in either arrested or aberrant development. Development is further distinguished by the fact that it always involves new organizations which have been synthesized to deal with newly discovered external or internal contingencies/information, and to integrate the new with pre-existing structures.

Within this sytem, development entails both the processes of *differentiation* and *integration*. Differentiation involves the creation of new and more specialized structures (whether these structures are characterized by functionally different neurotransmitter systems, regional and subspecialization within the frontal lobes, or structures of self-consciousness) each with different functions. Integration insures that development rather than being characterized by the autonomous functioning of highly differentiated structures, is distinguished by increasing interdependency between the evolving subsystems in the creation of a truly interactive system.

Typically, integrative activity is presumed to give rise to hierarchic organizations, with each layer in the hierarchy representing a different level of developmental achievement (i.e., lower levels represent acts of primitive organization, whereas . . .). Furthermore, higher levels are presumed to exert a regulatory influence upon lower levels. Failures to develop higher levels or disruptions of their normal functions give rise to disordered (i.e., unregulated) activity which in psychiatry would be evident as particular kinds of psychopathological conditions. Finally, among certain variants of an organic epistemology, one of which Harris has chosen, a dialectical motor is inserted to drive development through the resolution of conflict.

This, in short, is a brief sketch of the schemata employed by Dr. Harris. With it he has been able to organize and integrate recent findings in psychiatry, psychology, and the neurosciences. Furthermore, he has been able to provide the epigenetic perspective so essential to the development of his thesis. What emerges is a thought-provoking analysis of the relation of brain to the development of self, together with a splendid examination of the psychiatric vicissitudes of that journey.

In conclusion, I should add that an appreciation of the root metaphor that Dr. Harris has chosen and its particularization in psychoanalytic theory, usually requires a readership who have already developed an acquired taste for that world view. The uninitiated frequently find it too confusing/abstract/counterintuitive to stick with for very long. But I suspect that if Dr. Harris, like Aquinas, is able to strengthen the faith of those already committed, he will have made a significant contribution to a field desperately in need of reunification.

Jeff Rosen, Ph.D.

INTRODUCTION

In the days before the introduction of psychotropic drugs we used to think of psychopathological symptoms as disturbances in human nature. The wholesale use of psychotropics has impressed us all with the fact that disturbances in human nature take the form of biochemical disordering. The consequences of this impression have been (1) a salubrious recognition of the primacy of imbalance in some mental illness and (2) a wholly unsuitable idea that cases in which biochemical imbalance is a secondary feature may be "cured" by psychotropic means. Behind much contemporary puzzlement is an uncertainty about the rightness, as well as the possibility, of reconciling psychoanalytic findings and theory with new biological facts and theory as they have been produced for us by the brain scientists. What follows is an attempt to address, and I believe resolve, this dilemma as it faces all of us, psychiatrists, psychologists, neuroscientists, social workers, nurses, art and activity therapists, rehabilitation workers, administrators, educators, lawyers and judges, indeed the educated public at large. It begins with the fact that the primary origin of human conflict and suffering in social/familial life or in the life of the brain is a burning and unresolved issue for all mental health professionals.

The lack of certainty about where mental illness begins has left many of us with a lack of belief in our own ability to provide suitable help for the mentally ill. This profound identity dilemma affects all major areas of professional work. However unnecessarily divided they may all be, the divided beliefs we all experience lead to splits and inconsistencies in professional address. To give an example, the mental health field is interpenetrated with experts in chemotherapy. In an era of rapid expansion in this area, how is the less informed worker to avoid standing in awe of these specialists? Why, indeed, should he not be impressed? How is the worker to avoid idealizing or denigrating the purveyors of psychotropic medication? The medicators have seemed to dominate for years, and the social consequences of each new wave of psychotropic medication have been great. Their use of major tranquilizers (e.g., thorazine, haldol) emptied the state hospitals and put the patients back on the street until they ran out of medication. Then the antidepressants (elavil is one) gave hope that depression might be permanently alleviated. Finally, the antimanic, lithium, completed the therapeutic arsenal. Every patient may now be medicated into temporary quiescence. The double temptation facing the worker who finds years of therapy with a hypomanic patient superseded by lithium, is to feel the futility of his or her own brand of psychotherapy and to identify with the aggressor, the guru who has dispensed the chemical magic.

Pragmatically, one wonders every day if medication should be used with particular patients. The expert in emotional life knows that for the patient a belief in medication detracts from a belief in the efficacy of his or her own initiative. If the professional and the patient believe that the patient's conflict and pain are biologically caused and rooted, then the carefully established psychoanalytic transference pales as a therapeutic instrument. Doubt about the efficacy of psychotherapy surfaces under the scrutiny of peer review since the economics of treatment force the professional to justify therapeutic decisions.

The biologically minded must also ask themselves questions. In addition to the short and long range side effects of all drugs, we must consider the effect of chemotherapeutic intervention on human initiative. If the causes of personal suffering and conflict are ultimately biological, then emotional *process* must be a myth superimposed on the biological relationship. Another issue of great significance for the biologically minded is whether or not neuroleptic drugs destroy the process of recovery from emotional illness at the same time that they minimize symptoms.

The splintering among professionals in the mental health field has been exacerbated by decreased communication between groups and increased jargon within groups. Professional journals have proliferated to serve the communication needs and careers of entrenched minorities; often they fail to communicate to the field as a whole. Editors of book-length works no longer tend to feel that they are in a position to evaluate the content of new works, and therefore send works for review by a contingent that appears to support their own marketability biases. Marketability becomes a function of jargon expansion and of the biases of a particular splinter group that seeks to find new adherents.

Clinical Neuroscience is meant to repay the educated person's effort in reading it by providing a fieldwide perspective, in the light of which it will become far easier to evaluate diverse sets of information within this field. Cognitive, analytic, and neuropsychological approaches; Piagetian, Freudian, and Lurian adherents, are equally at home within the frame provided by *Clinical Neuroscience.*

A paradigm is the most general of frames for organizing operations, observations, and information within a defined field of inquiry. *Clinical Neuroscience* offers a paradigm of the brain/mind that harmonizes the major psychological and biological processes in a common structural context for studying human adult identity as that identity promotes problem-solving. *Clinical Neuroscience* is designed to address, and resolve, brain/mind consonance issues chapter by chapter as (1) didactic issues, (2) diagnostic issues, (3) and (4) causal issues, (5) theoretical issues, (6) developmental issues and (7) therapeutic issues. I have tried in this book to argue on the basis of an accumulated weight of evidence.

As Kuhn says in *The Structure of Scientific Revolutions* (Kuhn, Thomas, 1962) tremendous resistance meets fundamental changes in a scientific paradigm because such changes aim at overthrowing established patterns of experiencing and understanding the world. Only a sense of profound cognitive and emotional dissonance can lead an individual to re-examine the fundamentals of professional identity. The reader is therefore asked to suspend professional disbelief in order to test the capability of the structural paradigm presented in *Clinical Neuroscience* to reorient a portion of his or her scientific thinking. I do not claim that every detail used in support of the paradigm is correct; only that as far as I know every structural element in the paradigm, upon which the principles of the paradigm are based, violates no basic fact of any of the branches of neuroscience that the paradigm is meant to elucidate.

The *structural paradigm* itself is laid out in Chapter 1, as a necessary outgrowth of the salient neuroscience research of the past 20 years. I have attempted to choose the more classical articles, findings, and authors in the presentation of informations that are new, fundamental, and which must be accounted for in the paradigm. Although the paradigm is laid out in terms of neuroscience research findings, it could just as well be laid out in clinical terms. Ultimately it must be validated by its ability to clarify the individual clinical practitioner's experience with the whole gamut of emotionally disturbed patients.

The first principle of *Clinical Neuroscience* is that the brain works through *hierarchically organized functions.* The most superordinate of these functions is a social structuring one. Our very notion of a particular paradigm is enmeshed in our social expectations of our role in society within various institutions of society. *Clinical Neuroscience* holds that social functions themselves are the result of an interaction of evolving intrapsychic mind/brain structure with reality. A part of the brain, the prefrontal apparatus, has come to support the structuring of social processes. The same part of the brain structures the most superordinate of the mind/brain functions—the *identity functions.* The experience of intrapsychic identity depends on a set of brain/mind functions that are capable of unification as a mental system of consciousness. The intrapsychic superordinate level of (1) *Identity-functions* organizes a mental system of (2) *Ego-functions,* that, in turn, regulate a group of (3) *Neuropsychological-functions* that are intermediate and variably linked between the intrapsychic and neurological organization, and that finally regulate a group of (4) *Neurophysiological-functions* that are not directly accessible to the mental organization of consciousness.

The concept of variable linkage in the constitution of brain/mind, and indeed brain/brain neurological systems is basic to the structural paradigm. Functions are neither localized nor generalized in the brain. Instead they are hierarchically organized with an increasing range of lower level variable links that can take part in the system at a particular time. This concept of variable linkage explains why the higher cortical functions can be so stable that they can continue operationally despite the destruction of brain tissue that supports neuropsychological functions that are usually incorporated into the higher level functions. The concept of variable linkage is also useful in helping us to understand that the system-consciousness itself can be variably linked at the prefrontal level so that one or another particular structural form of consciousness can organize the whole, immediate, unified sense of intrapsychic identity.

An understanding of the principle of the functional hierarchy is of great importance in diagnosis. Nevertheless, the current diagnostic manual ignores this issue. *Diagnostic and Statistical Manual III* takes a phenomenological descriptive approach and does not distinguish the functional level of the phenomena under observation. Instead, it often requires the patient's conscious, introspectively derived report of pathological phenomena, yet it does not specify how close to the psychological or neurological level of integration a symptom may be (consider a low level of energy or impaired concentration). To remedy this confusion, Chapter 2 provides a methodological diagnostic approach based on a classification of *DSM III* symptoms and phenomenology according to the logical ordering of the functional hierarchy.

The functional hierarchy principle is also useful in providing a causal framework for understanding phenomenology and diagnosis. *DSM III* has been criticized for failing to bring cause into play in its descriptive guide to diagnostic conclusions. *Clinical Neuroscience* uses the functional hierarchy to propose a comprehensive theory of psychosis: psychosis occurs when social and identity function disruption leads to a deregulation of essential ego functions that are regulated by the particular higher level functions impaired. In this view, the person mired in psychosis suffers a continuous disruption in identity functions, both through increasing social and interpersonal discord, and also because losing integration, the lower level functions are disturbed. The experience of disturbed ego functions, such as loss of memory, time sense, concentration, or initiative leads the psychotic individual to abandon consciousness-organizing identity functions.

A second principle derived from the paradigm is that consciousness is synthesized as a product of a systematized, prefrontal, lateralized organization of four discrete zones of tertiary neurones. The *lateralization principle* is based on two anatomical facts. The first is that the medial zones of the prefrontal apparatus assess the limbic system, while the lateral zones of the prefrontal apparatus assess the sensorimotor analyzers. The second is that the two hemispheres have evolved a different mode of analyzing information so that the dominant hemisphere analyzes sequential information, while the nondominant hemisphere analyzes simultaneous groups of information. These facts make us aware that the zones of tertiary neurones that compose the prefrontal system-consciousness are in actuality four distinct zones: (1) a lateral dominant hemisphere zone (The Agent), (2) a medial dominant hemisphere zone (The Subject), (3) a medial nondominant hemisphere (The Self), and (4) a lateral nondominant hemisphere zone (The Object).

Clinical Neuroscience makes use of the lateralization principle to distinguish functional identity zones in terms that are clinically useful. This terminology derives from the vocabularies of neuroscience and psychoanalysis. To give an example of this language in use we may say on the basis of our knowledge of the general kinds of deficits that emerge with damage to each hemisphere that the dominant hemisphere integrates *pragmatic action* and *meaningful speech,* and at the higher prefrontal levels of integration organizes *goals* and *plans.* I have decided to call this overall form of consciousness *action-consciousness,* and the two identity functions that regulate it the *subjective-functions.* 'Subjective' is used here to indicate the sense of the center of action initiation. This is a gestural or expressive sense of identity as an initiating cause.

In contradistinction to the terminology of the action hemisphere with its subjective identity integration, and on the basis of our knowledge of the kinds of deficits that emerge with damage to the nondominant hemisphere, I have chosen to call nondominant hemisphere consciousness "reality-consciousness" and to refer to its identity functions as "objective" functions. The nondominant hemisphere organizes our sense of external reality as a spatial container that holds various discrete objects including our own bodies as sets of postural images. At a higher, prefrontal level of integration, the nondominant hemisphere organizes self and object representations. "Objective" is used here to indicate the sense of being an object in the world. The sense of participation in and harmony with, including one's affective relationship with reality, is subsumed in the objective sense of identity. I have chosen to call the integration of one's own social and perceivable self-in-the-world *Object-self.*

The lateralization principle has great importance in our understanding of the composition and cause of clinical syndromes. *Clinical Neuroscience* sees schizophrenia as a dominant hemisphere triggered syndrome in which the essential difficulty occurs with disruption and disintegration of the subjective identity functions. Analogously, *Clinical Neuroscience* sees the affective syndromes as nondominant hemisphere induced syndromes. In the affective syndromes the objective identity functions and the variably linked ego and neuropsychological functions are compromised. This view of syndrome formation and development leads to a greater surety of diagnosis within the framework provided by DSM III.

The third principle derived from the paradigm is a developmental or *epigenetic principle.* Twin processes, a dominant hemisphere triggered *growth process,* and a nondominant hemisphere triggered *adaptive process,* produce functional and structural changes in system-consciousness in a recognizable sequence of stages comprising the human developmental sequence. Chapters 3 and 4 of *Clinical Neuroscience* show that the disruption of stages in ordinary epigenesis are the causal links in the formation of pathological syndromes. Chapter 3 shows that disturbances in the sequence of stages that make up the psychological growth process give rise to such dominant hemisphere syndromes as schizophrenia and narcissism. Chapter 4 shows that disturbances in the sequence of change that makes up the adaptive process cause the nondominant hemisphere syndromes such as affective and borderline disorders.

Chapter 5 begins with a consideration of *trauma* and *fixation*. This chapter's corollary to the epigenetic principle defines trauma as a process of maladaptation, and fixation as a process of disturbed growth. The total context established so far allows *Clinical Neuroscience* in Chapter 5 to propose an integrated psychoanalytic theory based on mind/brain alignment. Here, the differences in the various psychoanalytic schools are seen to emphasize one or more of the brain functions described in the present paradigm. To give an example, Chapter 5 sees Kohut's theory of narcissistic development as describing distorted growth process and fixation in the dominant hemisphere, and it sees Kernberg's 'borderline disorder' as resulting from poor integration of nondominant hemisphere identity functions, through a maladaptive developmental process of the nondominant hemisphere.

Chapter 5 provides the reader with a developmental structural paradigm that gives evidence for the interactional process of brain and mind throughout the entire human developmental phases from earliest infancy to old age. Finally, Chapter 6 uses the paradigm to enhance the approach to treatment in both the psychotherapeutic and biological modalities. I have presented a variety of case histories to show how the principles of *Clinical Neuroscience* crystallize diagnosis, disclose the functional level of disruption, expose the relevance of trauma and fixation, and lead to a more complete and professionally assertive approach to interpretation that educates the patient.

Chapter 1

THE BRAIN IS THE ORGAN OF THE MIND

The Fundamental Concepts

The human brain has evolved a mental framework for human identity. The congruence between brain-system structure and mental structure can be approached through an understanding of four entities that include:

1. a basic form of consciousness that distributes itself between two hemispheres as a system-consciousness,
2. a drive system that is made up of three parts,
3. a regulatory system composed of two inhibitory systems and one facilitatory system, and
4. a hierarchically organized system of epigenetic layering.

THE TWO SPHERES OF CONSCIOUSNESS: BIOLOGICAL FRAMEWORK

R.W. Sperry received a Nobel prize for work that led to the discovery that the human brain contains two separate spheres of consciousness, one in each hemisphere (Sperry, R.W., 1968). He found that when the corpus callosum is cut the two hemispheres function independently, each unaware of the working consciousness maintained in the other. Since then we have learned a great deal about the characteristics of the two hemispheres. These characteristics contribute to an understanding of psychopathology. Much of what we have learned is best expressed in evolutionary terms.

The necessity of existing as a self-regulating entity in a pre-existing physical universe seems to have determined the evolution of the human two-hemisphere brain. One of the hemispheres specializes in the regulation of the internal environment. The other deals with assessment of the external world. I am going to call the first hemisphere, the dominant one, the *Action Hemisphere*. The second, or nondominant one, is the *Reality Hemisphere*. The Action Hemisphere regulates pragmatic action, speech, emotional expression, and consummation strivings through its ability to integrate a sequence of information. This hemisphere regulates impulse through a development of consciousness of subjective time, of future, of planning, and of goal formation. Insofar as it is the center of initiative I refer to its *subjective* integration of consciousness.

The Reality Hemisphere assesses the exterior world and forms images and representations of space. Its assessment function operates through the simultaneous processing of groups or gestalts of information. Each feature of information is accorded salience and combined with other features to form percepts—impressions of the world. This hemisphere maintains participatory process in the world through its assessment of the affective response to reality. The Reality Hemisphere produces images of the past as long-term memory experienced in the present. Insofar as it is the center of response I refer to its *objective* integration of consciousness.

The two hemispheres work together to form a unified system of consciousness. Flor-Henry (1978) describes the hemispheres as mutually inhibitory in their moment-to-moment mode of operation. I take this to mean that subjective and objective consciousness are reciprocal in their operations. Neuropsychologists have found, however, that all individuals show a predilection for one or the other of the hemispheric modes of consciousness. This probably genetic tendency accounts for predilections in psychopathology, character and style.

Three Drive Systems

The discovery of neurotransmitter systems and their neuropeptide modulators has revolutionized our understanding of neural systems. Before this discovery, neural systems could only be traced through direct anatomical means. Now the brain's major control operations can be followed through immunochemical radioligand detection. Thus, the observed distribution and characteristics of the major neurotransmitter systems allow for informed hypotheses about their role in the organization and maintenance of mental system. We will see that each sphere of consciousness is subserved by a major neurotransmission drive system, and that the coordinating system that holds the two hemispheres in common process is also subserved by a neurotransmitter regulating system of its own.

The Biological Framework of the Drive Systems

Despite the fact that there are many neurotransmitter systems, three of them—the norepinephrine, dopamine, and serotonin systems—are notable and distinctive because of the widespread brain communication and metabolic priming that each maintains. As neurotransmitter drive systems, norepinephrine, dopamine, and serotonin regulate the entire neural system from early in foetal development through epigenesis to old age. They determine patterns of neural growth and development; they maintain the energy systems' tone and trigger their physiological functions; they regulate the responsivity of the populations of neurones they service from moment to moment. These systems also modify their own action through a feedback autoregulation system so that their own neurotransmitter substance stops and starts, nourishes and triggers the very cell bodies that generate their own activity. Thus, each single norepinephrine neuron with its cell body in the locus coeruleus of the midbrain, and each single dopamine neuron with its cell body in the substantia nigra of the midbrain or the mesolimbic ventral tegmentum is an extensive brain system in its own right. Both norepinephrine and dopamine systems have ramifications that go down the spinal cord and determine sensory readiness, autonomic tone or motor reflex tone. Each system also has ramifications that extend up to the limbic cortex, the cerebellar cortex, and the neocortex.

Mountcastle (1979) describes the cortical network of norepinephrine influence in the following way:

> . . . from the two local points of entry, fibers pass tangentially to reach every cortical region and every cortical layer . . . the neocortex is seen to be interlaced in all directions by a web of fine noradrenergic fibers at 30–40 ug intervals . . . It appears that any single cell of the locus coeruleus may project widely upon the brain, including large areas of the neocortex, and sustains an immense and divergent axonal field. (p. 32)

Neurons of the serotonin system with their cell bodies in the raphé nucleus of the midbrain also course and ramify through the limbic system and the cerebral cortex. The fact is that these three neuronal neurotransmitter systems interact with each other at a variety of levels, including direct synaptic contact at the cell body levels and shared synaptic endplates that co-regulate receptor sites together with an array of neuromodulator peptides.

The evolution of differential functions in the two hemispheres has resulted in differentially active, parallel dopamine and norepinephrine systems in the two sides of the brain. For instance Glowinsky (1979) concludes that balanced differential activity of the two dopamine systems is a result of the different physiological activation of the two systems:

> (1) . . . the level of activity of the DA neurons seems to be inversely correlated to the extent of the dendritic release of DA in physiological states. (2) The asymmetric changes in the activity of the two DA pathways in response to unilateral sensory information provide further evidence for a specific role of the DA neuron in sensorimotor integration. (3) The marked opposite variations in the activities of the two DA systems could be initiated by messages preferentially delivered to one Substantia Nigra. In fact a balance exists between the activities of the two DA pathways. (pp. 1075–1076)

Thus, feedback activity in each system determines a differential laterality in those brain activities that are determined by the drive modulation of these neurotransmitter systems. The lateralizing distinction of right and left inheres in spatial orientation, perception, and motor planning.

In addition to maintaining lateralization, the neurotransmitter systems are direct modes of communication and control that regulate the functional hierarchy of brain systems. The functional hierarchy undergoes an epigenetic development that encompasses the evolutionary hierarchy of brain systems described by Maclean (1962), a reptilian kind of midbrain integration, a mammalian limbic brain integration, and a human neocortical prefrontal integration. This means that the dopamine, norepinephrine and serotonin systems collectively regulate each functional level of brain system integration.

It begins to appear (from the findings of recent radioactive ligand research through the immunotagging of neurotransmitters and an analysis of their receptor sites) that each of the major neurotransmitters operate through different forms of receptor binding in the postsynaptic synapse. One site triggers an adenylate, or related kind of cyclase enzyme to produce an energy increase in a monophosphate molecule. By (1) fitting into a protein configuration the neurotransmitter can enhance the energy status of the recipient cell, or by (2) fitting into a perhaps less specific binding site it can alternatively inhibit the energy enhancement. A third kind of energy effect relates to autofeedback, possibly on presynaptic sites that can act to 'downregulate' the metabolic activity of the neurotransmitter cell. For example, it appears that D1 energizes the dopamine system through its adenylate cyclase facilitation, while D2 may inhibit both co-regulated norepinephrine system facilitation and its own adenylate cyclase energy system (Creese, Ian, et al, 1983). Similarly, serotonin on its own may increase the energization of adenylate cyclase on its receptor cell membrane, whereas when it occupies a common site with dopamine, norepinephrine, or gabaminergic receptors it may act to inhibit or downregulate the receptor cell.

We must conceive of the neurotransmitter systems working differently at each hierarchical evolutionary level. Thus, as prefrontal cortical systems demand more energy for problem-solving, the norepinephrine, dopamine, and serotonin systems must amplify and comodulate the energy availability to the cortex. The process of amplification is partially dependent on neuromodulator feedback. When the postsynaptic regulators of the cortical sites are chronically upregulated they become supersensitive, finally losing their metabolic flexibility. As demands continue, therefore, a mechanism of stabilization has developed so that upper level energy resources do not become exhausted.

Neuromodulation of each neurotransmitter system is accomplished through diurnal homeostatic mechanisms, and in stress, through extensions of diurnal modulation that downregulates the neurotransmitter systems as they become overused. In line with the evolutionary development of neural system regulation, down regulation is most effectively accomplished at the cell-body midbrain level, while the limbic-hypothalamic level is best situated to modulate flexibility through a combination of up and down regulatory influences on postsynaptic receptor configurations. Neuromodulator and drive systems all come to bear on the PVH nucleus of the hypothalamus, which is the "visceral effector nucleus" that regulates the autonomic system interaction with higher levels (Swenson & Sawchuk, 1983).

In one sense the pituitary is the rostral extension of the gut. Modulatory peptides are present both in secretory glands of the gut and in the pituitary. A whole range of neuromodulatory peptides is also present in the pituitary, in the hypothalamus, in cell groups surrounding the hypothalamus, in limbic cell groups such as in the amygdala, and even in cortical sites. Consummatory pressures tend to be brought to bear through the gut-pituitary limbic axis, and dopamine plays a major part in the neurotransmission of these pressures. Thus, the lateral hypothalamic "feeding center" receives dopaminergic stimulation, while the medial hypothalamic 'satiety center' receives inhibitory adrenergic stimulation. The hypothalamus is a crossroads of different modulatory effects.

We may conceive of two generalized systems with extended neuromodulatory and neurophysiological components that are eventually and hierarchically regulated at a prefrontal level. The first is the consummatory dopamine facilitated system and the second is the pain system. The pain system is involved with response to reality pressures. Its extended neuromodulatory component includes the endogenous opioids and the ACTH system. This system is most responsive to the state of the noradrenergic neurotransmission. Because of the fact that all of the major neurotransmitter systems can exist as energy system amplifiers or inhibitors, because serotonin often mediates the effects of the other systems, and because all three systems are often involved in the regulation of a particular physiological function, we must be careful not to generalize beyond the notion of an action-consummation system, a pain-reality system, and a neutralizing intermediate system.

Evelyn Satinoff (1982) discusses the fact that physiological drive systems such as sex and thermoregulation are hierarchically regulated within a classical Jacksonian framework, with neocortical influences on top, followed by hypothalamic influences which are hierarchically superior to lower levels of integration.

> . . . In a Jacksonian framework, the hypothalamus is near the top of a hierarchy, directing the activities of thermoregulatory systems lower down. These other systems are capable of independent action, but without the hypothalamus they are much less effective. With diencephalic direction, appropriate reflexes and be-

> haviors are activated promptly and inappropriate ones are suppressed. The hypothalamus is near, not at the top of the hierarchy, because more encephalized mesocortical and neocortical systems influence the activity of the hypothalamus in temperature regulation. (p. 220)

It follows that individual neurotransmitter systems interact with brain structure systems at each level of the control hierarchy for those neurophysiological regulatory systems that we associate with drives.

The other hierarchy of systems that bears on our understanding of the relation of the neurotransmitter systems to drive in our paradigm is the hierarchy of prefrontal controls that develops over the span of an entire life. I call this lifetime unfolding of consciousness *epigenesis*. The neurotransmitter systems develop an ever higher level of cortical integration as new sources of motivation come on stream. The lateralization of higher identity functions tends to force the neurotransmitter systems into functional differentiation.

The general instinctual-biological effects of these systems therefore leads us to conclude that the dopamine neurotransmitter drive system constitutes an *action-drive* for regulating consciousness in the action hemisphere. It is responsible for regulating an action tone and for regulating a level of drive for consummation of all the physiological necessities of the internal milieu. It is likely that the norepinephrine neurotransmitter drive system also maintains the organization of consciousness in the reality hemisphere. It constitutes a *reality-drive* responsible for the impetus to assess and direct perceptual processing of the indications of salience in reality. Indeed, it seems reasonable to conclude that the serotonin neurotransmitter system constitutes a coordinating drive system *(a neutralizing drive)* responsible for maintaining a balance between the operations of the two catecholamine systems at each level of the control hierarchy of integrated brain functions. This means that the serotonin system is responsible for selectively inhibiting (neutralizing) the other two drives, thus distributing functional activities.

Psychological Framework of the Drives

The dopamine, norepinephrine, and serotonin systems correspond particularly to the libidinal, aggressive, and neutralizing drives of Freudian psychology. This is to say that dopamine presence readies system-consciousness to initiate consummatory action. The dopamine system may be equated with Freud's *libido* to the extent that *libido* refers to a developmental range of activities and sublimations, all of a consummatory nature. Study of the neurotransmitter systems, however, suggests that sexual activities and subliminations are simply one perceived face of a neurological drive conditioned by dopamine that includes a biologically initiated range of behaviors, ranging from the quenching of thirst to the desire to have one's expressions of love understood by the beloved. Because the drive is not purely sexual, it seems best to call it the *action-drive.* Similarly, norepinephrine may be equated with the aggressive drive insofar as aggression may be newly defined as an imperative to explore reality, both physically and perceptually for the sake of salient feature detection. When the individual finds novelty or threatening features in reality rather than familiarity, he or she may develop aggressive affects ranging from irritability to outright anger that tends toward the obliteration of reality features. Equating aggression with all forms and motives for behavioral hostility is misleading. For reasons that will be explained, this norepinephrine-fueled drive excludes such instinctual emotional expressions as rage, fight, and hatred—commonly associated with the word *aggression.* In this paradigm the norepinephrine drive will be called the *reality-drive.*

The action-consummation and reality-pain systems exert neurological pressures from various subcortical sources on the drive input to the prefrontal cortex system-consciousness. We might say that the latter system must selectively damp these dopamine and norepinephrine fueled drives according to its own synthetic function. In this paradigm, we will call the monitoring agency the neutralizing agency. The neurotransmitter serotonin accomplishes this neutralization.

System-Consciousness and its Distribution

A separate system operates, then, to maintain the highest level of synthesis of the brain's hierarchy in the prefrontal areas of the neocortex. In an editorial in *Biological Psychiatry*, Itil (1983) argues for the existence of such a system when he points out that *Nootropic* drugs only effect high-level brain functioning. These drugs strengthen the integrative brain rhythms and so help to stabilize high level brain functions. A related argument for the existence of an extremely stable, autonomous system-consciousness in the prefrontal lateral and medial tertiary neocortex is that the highest level of psychological functions are strongly resistant to destabilization. Indeed, it appears that these prefrontal areas maintain so stable a neural system that the higher psychological functions tend to maintain themselves through alternate distribution if a portion of their neural system is removed.

Thus, Mountcastle (1979) describes the general neural basis for systematized consciousness as

> perceiving, remembering, thinking, calculating, formulating plans for current and future action and consciousness itself . . . as the internally experienced and sometimes externally observable behavioral events produced by the ensemble action of large populations of the forebrain, organized into complex interacting systems. (p. 21)

The populations he refers to are massive groups of neurons in the frontal and prefrontal areas that work cohesively with one another. Luria (1980) has pointed out that large neuronal populations constitute many separate zones capable of doing communal work. These zones are surrounded by other zones that modify their working. The prefrontal system of the cortex is thus composed of zones that have evolved from diverse regions of the brain, but that are so refined and so late in their evolutionary appearance that they have evolved a synthesis as a single system. In Luria's words

> The prefrontal division of the cerebral cortex are situated anteriorly to the motor area (area 4) and premotor (areas 6 and 8) areas and comprise a series of formations (areas 9, 10, 11 and 46) situated on the convex portion of the mediobasal surface of the frontal lobe . . . anatomists now regard this region as one of the most phylogenetically recent tertiary regions; it possesses the finest structure and displays the most varied connections with all the remaining structures of the cerebral hemispheres . . . (p. 258)

Luria points out that the medial portions of the prefrontal areas regulate and assess the limbic system, while the more lateral portions of the prefrontal zones regulate and assess sensorimotor processes that are tied into cognitive functioning. Given the fact that there are two hemispheres, each with a medial and a lateral prefrontal population of neurons, we must think of the system-consciousness as containing *four separate zones.* This means that each of the four zones of system-consciousness maintains a vast *distributed* cortical system which it integrates, as well as various subcortical distributions with which it has two-way feedback relationships. Each zone of system-consciousness controls, then, a population of subzones of other neural populations that may or may not be included in the distributed system of the prefrontal zone during the performance of any particular system-consciousness function. The existence of distributed neurological systems is fundamental to our paradigm.

The concept of distributed neurosystems with multiple or flexible capacities is of basic importance in modern neurology. In one basic paper, Mountcastle (1979) has described the manner of intercollaboration of various related populations of neurons in the parietal lobe:

> It is my proposition that each of these classes of columnar sets of the parietal cortex is related by specific extrinsic connections to similarly segregated sets of modules in other cortical regions and in subcortical nuclei as well, and that these closely interconnected modular sets of different large brain entities form precisely connected, distributed systems, serving distributed functions. (p. 32)

From this it seems likely that the same mechanism describes the intercollaboration of the prefrontal zones. In another format, Bechtereva (1978) states the distribution thesis as it pertains to the prefrontal system:

> . . . the maintenance of human mental activity by a cortical-subcortical, structural-functional system with links of different degrees of rigidity . . . offers a basis for interpretation of the physiological mechanisms of mental phenomena, particularly the structural-functional aspects of their maintenance. Our investigations have shown that flexible links predominate in the system of cerebral control of mental activity. (p. 135)

The flexible links distributed among the four prefrontal zones of the system-consciousness are responsible for a brain-congruent, intrapsychic system that functions with alternative forms of organized consciousness.

One further anatomical subsystem of system-consciousness deserves mention. The premotor system has evolved frontal supplementary areas for speech and pragmatic movement. Both of these areas are involved in the selective prefrontal inhibition through which voluntary speech and action are carried out. Thus, both SMAs (supplementary motor areas) must be considered to be within the distributed system of the dominant hemisphere lateral prefrontal zone of system-consciousness. According to Eccles (1982)

> the SMA represents the liaison area of the brain for intentions and . . . the dynamic patterns of its activity carry the information

> derived from the intentions and relate it to the repertoire of motor programs that are brought into action by outputs from the SMA . . . (p. 282)

Interestingly, both SMAs come into play when a voluntary movement is carried out.

The Psychological Framework for System-Consciousness

> Preliminary integration of all stimuli reaching the organism and the attachment of informative or regulatory significance to some of this—the formation of the "provisional basis of action," and the creation of complex programs of behavior; the constant monitoring of the performance of these programs and the checking of behavior with comparison of actions performed and the original plans; the provision of a system of feedback on the basis of which complex forms of behavior are regulated—all these phenomena in man take place with the intimate participation of the frontal lobes, and they account for the exceptionally important place of the frontal lobes in the general organization of behavior. (Luria, 1980, p. 248)

Thus Luria indicates the specificity of ego-functioning maintained by the frontal lobes. In our terms, Luria's description means that the lateral dominant prefrontal zone of system-consciousness proposes an action plan. The lateral nondominant prefrontal zone assesses its progress. The medial zones monitor and regulate the working collaboration.

At the hierarchically topmost level of system-consciousness each of the four brain functions is at once psychological and neurological. I call these zonal functions *identity functions,* because when the four functions work together they form essential components of the psychological synthesis of identity while they account for the specifically human experience of intrapsychic consciousness. A proviso to be added here is that the lateralization of identity functions about to be described does not apply point by point to the *ego functions,* which are subordinate to the identity functions. The ego functions (I will show later) are organized bihemispherically. Let us consider the roles of each identity function.

I call the first of these identity functions *The Subject.* The subject is the dominant hemisphere medial prefrontal zone of integration of consummation tendencies. It contains the urge for expressive communication and fulfillment. The subject consciousness *wants.* The second function is *The Agent.* The agent is the dominant hemisphere's lateral prefrontal zone. It integrates the generation of meaningful speech and pragmatic action and contains the verbally generated sense of conscious purpose. The agent consciousness *wills. The Self* is the third identity function. It resides in the nondominant hemisphere medial zone. The self conveys an affective consciousness that deals with the progress of action plans in reality. The self-consciousness *responds.* The last function is *The Object.* It composes perceptual representation of reality and objects within reality including other persons and the *object-self* as a central object that participates in the world. By object-self I mean the mental representation of the self in the world. The object consciousness *surveys.*

The social and biological necessities demand that in each phase of life the individual remake identity functions. The superseded functions continue to have a structural role in identity that is hierarchically subordinated to the current ones. They become absorbed into *ego-functions,* which is to say that they form a repertory of ways of wanting, willing, responding, and surveying. To give an example: the adult function of agency controls a hierarchy of ego functions that includes the generation of meaningful speech (from the eighteen-month period), goal formation (about six years), and planning (which is an adult function). Following the distributed ego functions into their neuropsychological components, we can say that *neuropsychological functions* comprise the subsidiary functions that are contained within the ego-functions. Thus, to analyze the ego-function of speech generation, the component functions of (1) premotor dominant hemisphere *morpheme* sequencing and (2) parieto-temporal inhibition of irrelevant sounds in *phoneme* constitution are both neuropsychological functions that depend on the intactness of particular brain areas for their continuous operation. The ego-function of speech generation that combines these two neuropsychological functions into meaning sequence can continue to have functional intactness even if one or both of the neuropsychological components are damaged.

A consideration of the medial zones of functional identity integration expands our notion of the functional hierarchy to include *neurophysiological functions* within the distributed system of consciousness. The consummatory pressures that give rise to expressive tendencies and specific consummatory motives, and the painful pressures that give rise to affective motives, are the cortical derivatives of extended neurophysiological systems with their roots as far away as the guts, glands, and autonomic nervous system. The selective absorption of amino acid precursors of the drive systems in the gut, and the hormonal and autonomic activation of hypothalamic and other neural sites, determines the ultimate kind and intensity of drive pressures relayed to the medial identity zones.

One (Acetylcholine) Facilitatory System, Two (GABA) Inhibitory Systems

Thus far in the construction of a brain/mind paradigm we have posited the interaction of two hemispheres with three drive systems inside an overall regulatory system consisting of four zones of integration. We must now add to the paradigm the effects of a facilitating acetylcholine system that ascends through the reticular activating system and through the reticular portion of the thalamus to activate the cortex. This central core of facilitation is responsible for wakefulness and attention. The cortex also uses glycine as a facilitory neurotransmitter. Serotonin transmission enhances centrally active gabaminergic receptor inhibition in prefrontal control (Leysen, 1984). Serotonin-gabaminergic dual inhibitory regulation of the frontal cortex provides the means by which selective inhibition determines working cortical zones. Overall, the serotonin system is in a mutually inhibitory regulatory relationship with the acetylcholine system so that selective inhibition yields selective attention, while at its widest distribution the serotonin system shuts down consciousness and so produces sleep onset and slow wave sleep. Specific phasic activation of the cortex by the acetylcholine system proceeds working zone by working zone through a feedback arrangement that continuously reactivates cortical-thalamic distributed segments. You might say that the prefrontal system-consciousness is a master switchboard of consciousness. The reticular portion of the acetylcholine system turns the system on.

Once the switchboard has been turned on, two inhibitory systems—one for each hemisphere—must form an alliance with the dopamine and norepinephrine systems in determining whether cortical programs will go on to completion. These two inhibitory systems are among the widespread inhibitory gamma amino butyric acid systems (GABA-gabaminergic). Indeed, there are many gabaminergic systems in the brain, and their mechanism of shutting chloride channels, and thereby reducing potential activation of neurons is a powerful inhibitory process. The bilateral GABA inhibitory systems comprise the mechanism through which anxiety deactivates programs which are going awry.

In psychological terms the dual anxiety systems are made up of *error-anxiety* in the dominant hemisphere and *novelty-anxiety* in the nondominant hemisphere. This is clear from an inspection of the mechanism of the two major anxiety systems. One inhibits wrong actions and is therefore related to dominant hemisphere functioning. The other inhibits further input of perception into consciousness of reality when novelty is encountered. It is therefore a nondominant hemisphere-related anxiety system. Bechtereva (1978) describes the entire anxiety system (which she calls an error-detection system) as hierarchically controlled:

> The presence in the brain of neuronal populations that react subtly to an incorrect action, as well as the different degrees of the intensity of the reactions to an error in a psychological test, leads to the assumption that the different elements of the error-detection apparatus are not equally significant and that they are arranged in a hierarchy . . . in the realization of a mental activity, the error detectors of the neuronal populations of the nucleus caudatus are of great importance. (p. 135)

Recently, a great deal of work has demonstrated that the gabaminergic systems are stabilized through the use of benzodiazepine drugs. These drugs have been found to bind most densely to gabaminergic sites in the cortex and in certain subcortical limbic sites such as the amygdala (Saric, 1982). Central GABA receptors are highly concentrated in the cortex and in the limbic system, having increased in scope and extent during mammalian evolution (Richards & Mohler, 1984). GABA inhibition appears to have evolved as a hierarchical control mechanism.

Error and novelty detection relate to a process for comparing the unfolding of present experience with expectations based on its past unfolding. According to Luria (1980) the hippocampal cortex of the limbic system contains populations of neurons that respond to 'novelty.' The limbic cortex organizes long-term emotional memory. When discordance with present experience occurs, the subcortical nuclei of the limbic system, the septal amd amygdalar nuclei, give rise to phasic stimulation that interrupts system-consciousness. Each amygdala and each septum has balanced facilitatory and inhibitory regulation of cortical-directed output. The inhibitory regulation appears to be at least partially effected by a cortical feedback maintained by the gabaminergic systems. Either release of inhibition or increased limbic pressure can lead to discharge phenomena, including the activation of of instinctual behaviors that lead to gratification, or to fight or flight activities. It therefore appears that the two gabaminergic systems stabilize system-consciousness operations. It seems, moreover, that they themselves develop as hierarchical control systems corresponding to the hierarchical control that develops in system-consciousness through epigenesis. Limbic signals can evoke an examination of memory, or, when prolonged, determine instinctual responses that circumvent system-consciousness.

Research shows that there are indeed two gabaminergic anxiety systems that interact with all three of the major neurotransmitter drive systems. Saric (1982) points out that there appear to be two major anxiety systems. He posits one norepinephrine responsive system that leads from the locus coeruleus to the septum and the amygdala. This system responds to outside conditions of novelty and gives rise to the fear response. He posits a second system related to muscular tension. He believes that the muscle-relaxing properties of the benzodiazepine drugs relate to this system. What this means for our paradigm is that the novelty-system responds to the norepinephrine system and to processing of information from the outside world. This system mediates flight-fear reaction: the second or error-anxiety system fits into our paradigm scheme as a dopamine related gabaminergic system that operates when error tension occurs, and which effects the musculature to oppose actions that may turn out to be erroneous. This anxiety system must mediate fight-rage response when the anxiety system produces insufficient inhibition of consummation strivings.

On occasions when repeated novelty or error signals are prolonged, the catecholamine systems, serotonin system, and GABA systems can no longer maintain the functional coherence of system-consciousness. According to Stevens (1979), who discusses this point in her description of the opposition between psychosis and epilepsy:

> There is evidence that the catecholamine systems exert a powerful inhibitory effect on propagation of EEG spike activity and may thus limit excitation to specific regions of the brain in which the spike mode of signal transmission is physiologic. Neuroleptic agents that block catecholamine receptors or lesions of the catecholamine pathways induce epileptiform activity on the EEG and lower the threshold for propagation of convulsive discharge. (p. 260)

Stevens points out that the excitation discharges emanate from the amygdala and septal areas of the limbic system. In addition to the major regulations the balance of facilitation and inhibition in the limbic system is also determined by a facilitating glutamate and an inhibitory adenosine system as well as various neuromodulators.

The foregoing allows us to hypothesize that in epigenesis layers of system-consciousness alternate with layers of defensive inhibition. The ordinary anxiety system that can inhibit large portions of system-consciousness comes to function continuously when a new system-consciousness layer develops epigenetically. In epigenesis, the anxiety system thus forms a set of semipermanent barriers that psychoanalysis calls primal repression (at eighteen months), reaction formation (at thirty months), and repression proper (at forty-eight months). The brain maintains access to earlier layers of consciousness through hierarchical control of the defensive system; for creative work, psychopathology, the dream process, and the psychoanalytic process all show that under some circumstances these hierarchical inhibitions can be turned off.

This means that these anxiety-inhibitory systems, in conjunction with the inhibitory control produced by the selectivity of the interaction of the three drive systems, form a prefrontal system-consciousness that is basically a system for maintaining hierarchical control and selective regulation of activating processes. When that overall control is threatened, the consequences are either seizure discharge, or such instinctualized behaviors as the dominant hemisphere regulated flight-fear reaction. Ordinary error and novelty inhibitions maintain the control of system-consciousness over these overly intense facilitatory processes.

The Epigenesis of Identity

Epigenesis builds up hierarchical layers of system-consciousness. A layer forms when the unfolding of a biological determinant exerts an effect on the whole system of identity stabilizing, or when a necessary and sufficient experience in reality promotes the formation of a new structure for experience. The stratification of identity integers and identity systems gives rise to ego-functions that act as intermediary links in the newly formed functions of the identity system. Luria (1980) emphasizes these points as necessary to the understanding of the neuropsychopathology that develops due to lesions at different points of life:

> . . . the relationships between the individual components of the higher mental functions do not remain the same during suc-

> cessive stages of development. In the early stages, relatively simple sensory processes, which are the foundation for the higher mental functions play a decisive role; during subsequent stages, when the higher mental functions are being formed, this leading role passes to more complex systems of connections that develop on the basis of speech, and these systems begin to determine the whole structure of the higher mental processes. . . . This concept implies that the character of the cortical intercentral relationships does not remain the same at different stages. (p 35)

Cortical intercentral relations refer to the identity zone interactions. The two medial zones together regulate the global ego-function *emotion*. The two lateral zones intercollaborate to maintain the global ego-function *cognition*. The two dominant hemisphere zones together *propose* action plans, while the two nondominant hemisphere zones together *assess* their progress.

A consideration of the development of system-consciousness at different life stages requires study of both emotional-medial development and cognitive-lateral development. Medial development accrues when new drive pressures are relayed to the prefrontal cortex. Because intrapsychic awareness of identity develops vis-à-vis the emotionally significant object relationship in each epigenetic period we must consider each stage of emotional medial development in terms of its instinctual drive mandated form of *object-relations*. Cognition gels in each stage as a means of intrapsychic communication of consciousness capable of combining salient reality assessment with developmentally stage-appropriate action-plans. Cognition is the intrapsychic mode of synthesizing subjective and objective informations in a mental system in a form that can stand for the whole activity of the system-consciousness.

Paradigm Applications to Human Nature

The problem-solving work of system-consciousness throughout epigenesis is the attempt to produce an identity between subjective necessity and what is objectively available. When it is accomplished this achievement always brings about (1) a feeling of resolution in the state of intrapsychic identity, (2) a relaxation that gives pleasure, and (3) the end of the action program in question. To be sure, the nature of what is sought develops and changes during epigenesis, so that the various programs for proposed actions and satisfiers change.

As life goes on, the hierarchical development of system conscious-

Table 1
Human Epigenesis

Stage of Life	*Biological Development*	*Emotional Development*	*Cognitive Development*
Early Infancy	System-Consciousness Coalesces	Subject-Object Mirroring	Imagery Formation
0–2 months	Autonomic Regulation	Instinctual Relating	Image Intensity Control
2–6 months	Frontal Regulation	State Sequencing	Memory Facilitation
6–18 months	Secondary Zones Mature	Subject and Agent	Transitional-Object
Later Infancy 18–24 months	Secondary Zones Combine	Self and Object	Verbal Generation
Early Childhood 24–42 months	Tertiary Cortical Development	Subject, Agent, Self, Object Differentiation	Categorical Representation
Oedipal 42 months to 6 years	Neocortex Myelinates	Subjective Egocentrism	Symbolism
Latency 6–11 years	Neocortex Differentiates	Objective Socialization	Abstraction
Puberty 11–14 years	Hormone System expands	Reflective Egocentrism	Preconceptualization
Mid-adolescence 14–18 years	Hormonal modulation	Reflective Narcissism	Subjective Conceptualization
Late Adolescence 18–21 years	Hormone up-regulation	Reflective Relatedness	Objective Conceptualization
Adulthood 21–40 years	Neural/Hormonal Homeostasis	Subjective Social Relatedness	Subjective problem-solving
Maturity 40–70 years	Neural/Hormonal down-regulation	Objective Social Relatedness	Objective problem-solving
Old age 70 years and over	Neuronal/hormonal deterioration	Universal relating	Hierarchical Generalization

ness proceeds like weaving. First one hemisphere's consciousness and identity integers lead development. Then the other hemisphere leads. We can see, in looking at the course of human life, that a subjective period is always followed by an objective one. In the terms of the paradigm we have been establishing for the neurosciences, this means that the dopamine regulation increases in scope and frequency of usage first. Norepinephrine control expands in the period that follows. We must assume, then, that as the system-consciousness develops its hierarchical control, synthesis, and regulation, all of the neurotransmitter systems must gradually increase in scope and availability. We also see that as the demands of a particular neurotransmitter increase, a basis must be provided not only for replenishing supplies, but also for a targeted replenishment so that precisely those areas that use up their neurotransmitter during the course of conscious process are replenished.

At day's end we review our accomplishments. We see how far we have gone in implementing fundamental identity-syntonic plans. In the morning, we awake with a subjective recognition of our identity: we are about to set out on a quest to satisfy the day's action-plans. Thus, we renew our contact with essential identity-syntonic designs at the beginning and at the end of the day. By night, when we forego further accomplishment, we see that the available energy resources have been depleted. We must not expect too much from ourselves. We have learned, and we are reminded by watching our children, that the coming of night means the loss of ability to neutralize intentions. Children especially show more primitive behavior at night. The sequence often begins with an impetus to intensify, that is followed by a loss of personal control, that is then replaced by parental control. We recall the naps we took as children, begun after a rage at relinquishing our action wishes. These phenomena relate to the economy of our neurotransmitters and neuromodulators. Gradually, during the day, our norepinephrine usage increases until it peaks at midnight. Then our serotonin system must function to put us to sleep by producing a blanket inhibition of our cortex at sleep time. In our dreams, it is possible, precisely the cortical areas depleted by irresolute conflicts are targets of metabolic replenishment, for the locus coeruleus fires away its putative metabolic energization as the REM state ensues.

Recall the experience of learning new functions in childhood. First, a great effort and concentration goes on, sometimes for days. We see our children struggling, irritably, until one day they master the new activity. At this point a new psychic economy emerges. The energy that has been previously expended to build the activity is now available for its enjoyment. I maintain that such phenomena are the outgrowth of the development of the neurotransmitter systems and their sequelae—the new economies that are introduced with new modes of problem-solving.

Functional Processes

Our neuroscience paradigm must be expanded to account for ordinary psychology. Basic Freudian and cognitive psychology, reinforced by Luria's work, has it that the psychic apparatus functions by proposing satisfaction as a sequence of actions, and by positing a set of conditions that must be fulfilled to produce the satisfaction desired. The coordinated prefrontal system must render a judgment that the object of satisfaction is "good' (that it satisfies subjective requirements to reciprocate motor discharge), and that it is also "true" (that it has all the familiar features that make it verifiable). The essential point, then, is that judgment finds *salience* is the object. *Salience* is a combination of meaning (which is a product of sequencing) and familiarity (which is a product of simultaneous comparison). Salience requires limbic organized memory-function to signal *error* or *novelty*. No matter how complicated a problem of adult life may be, it must always be solved through a series of propositions and proofs, each facilitating the next, and each organized as a miniature version of the whole. The experience of identity integrity depends on the successful accomplishment of such coordinated tasks.

What happens to the functional process when necessary satisfaction *cannot* be attained? Failure of problem-solving leads to a sequence of automatic corrections. The corrections involve assessments of the action program, which may have been improperly sequenced. They also involve assessment of the presence of novelty in the features of the salient object that has been selected as the means of satisfaction of the action program. These assessments interrupt the smooth course of thought. The interruption is experienced either as an anxiety signal about the proposed action of speech, which is then blocked—or else the interruption shows itself as a signal that there may be danger in the environment. The interruption to the psychic apparatus initiates a *reorientation*. This is a reflex inhibition of the system-consciousness prefrontal apparatus accompanied by new information, either about the proposed sequence or about reality. Once the new information is evaluated it is reconceptualized as part of the ongoing action program. This may lead to a possible midcourse alteration of the action program, for in the last stage of an action program the synthetic function of judgment assesses the likelihood of the plan's probable outcome of enhancing future identity functions.

If error and novelty anxiety still fail to supply course correction, and if the synthetic function finds no threat to ongoing identity functions either through failure or through damaging consequences, then the system-consciousness invokes fantasy and/or long term memory to solve the ongoing problem. Fantasy, of course, is an interweaving of solutions stored in long term memory with projections of a desired future.

Long term memory is itself organized generically, according to life events that led to pleasure or pain in the past. Affective response as prolonged frustration leads to the substitution of long term memory or fantasy for experience of the present reality. It is a notable feature of this process that the experience of personal identity heightens markedly during reviews of long term memory and of fantasy creation, for the person engaged in remembering or fantasizing must intensely cathect his or her present identity in order to safely re-experience its past organizations. In this regard the fantasy or memory state must share some common elements with the dream state. Certainly, during the ongoing action program the process of judgment maintains an observing stance during the introduction of long term action memory and of long term experiential memory into the purview of system-consciousness. If the long term information is capable of intersecting the present experience in such a way that action patterns or features can be reanalyzed, then the action program may be remodelled and benefit from the excursion into the past.

If, however, the invocation of the past fails to contribute to the remodelling of the action plan, and if anxiety signals mount from signals to discharge, then further interrupting processes occur. After reorientation, instinctualization of behavior may take the place of further system-consciousness processing. These are fight or flight behaviors (originally described by Walter Cannon) instituted by the catecholamine system. Thus, we may now add to our paradigm the idea that fight or rage reaction is a sympathetic nervous system response to a prolonged or repeated error signal. Essentially, fighting is an instinctual attempt to wrest satisfaction from a situation when vital consummation plans are blocked. This means that the fight response and fight anxiety are both dominant hemisphere modes and might be called action-anxieties in that their regulation derives from action-drive and from the action-hemisphere. For this reason this paradigm separates fight and rage from aggression which is here defined as a function of the nondominant hemisphere. Conversely, fear or the "flight reaction" may be seen as a response to prolonged and dangerous novelty. This may take the form of unexpected reality response to action plans or the eruption of traumatic and unforeseen reality. The flight response, then, is an instinctual behavior that effectively removes the surrounding reality.

When they supervene, fight or flight unleash chains of neurophysiological events. When, however, simple flight and fight are ineffective or impossible, and when the anxiety continues then a *kindling phenomenon* begins (Goddard, 1969). This is the beginning of a change in the operation of system-consciousness. We must add to our neuroscience paradigm at this point, that kindling is a limbic subcortical response to repeated stimulation beyond the capacity of system-consciousness to regulate. Kindling produces imperative changes in the structures of consciousness. In this phenomenon, the amygdala and the septal nuclei begin prolonged spike responses that drive the hippocampus to take in new experience and to form prolonged memory. According to McGeer, Eccles, and McGeer (1978):

> . . . with successive repetitions of this mild stimulation about every half hour, there was a progressive increase in the potentiation so that after the fifth there was an enormous potentiation continued for three hours . . . In chronic experiments with implanted electrodes a similar potentiation was observed even for several weeks after conditioning by six brief trains of stimulation . . . we can conclude that in these experiments it has been demonstrated that the spine synapses on the dendrites of the hippocampal granule cells are modifiable to a high degree and exhibit a prolonged potentiation that could be the physiological expression of the memory process. (p. 504)

New life conditions precipitate the formation of new generic memories. Indeed, prolonged and repetitive new experience gives rise to new neural and psychological structures that must be accommodated by changes in the identity integrations of system-consciousness. Essentially, these changes are made either on the basis of *fixation*, which may be defined as unexpected profound pleasure, or on the basis of *trauma*, which is to say on the basis of unexpected and profound pain. Both fixation and trauma lead to kindling phenomena and consequent brain-mind change.

Soto (1983) found that once kindling was begun in the cat it continued for many days, generating repeated stereotyped behaviors that were maintained by the dopamine system. Blocking the dopamine system blocked the stereotypic behavior. The stereotypic behavior in its turn was proof against the onset of seizures. Thus, we see that the liberation of consummatory instinctual behaviors occurs in a situation which may be taken as a model for fixation.

Adamac and Adamac (1983) extend the definition of kindling to include all temporally spaced neurobehaviorally active agents that mediate lasting change in brain function. They assert that GABA is a most effective agent in blocking kindling phenomena. We see from this why benzodiazepines (GABA strengthening agents) are effective in blocking seizures induced by the limbic system. The general implication of this line of reasoning is that the anxiety systems inhibit the formation of kindling responses, and it must be that only when the anxiety systems have become ineffective or exhausted that kindling as a brain changing process can begin.

Kindling is a process that occurs particularly in those brain areas that give rise to the onset of seizures most easily, and which are also implicated in the formation of new long term memory. Thus, according to Rosen (1984):

> . . . The term kindling refers to the phenomenon whereby periodic electrical stimulation of the amygdala or hippocampus at levels of current initially too low to produce a behavioral response, ultimately results in a progressive development of seizures through a number of distinct stages . . . kindling modifies the function of pre-existing neuronal networks, particularly within the limbic and mesencephalic reticular systems . . . This phenomemon has been associated with the important function of the hippocampus in memory consolidation. . . (p. 29)

Thus, the hypothesis is warranted that the failure of system-consciousness to prevent recurrent novelty with intense limbic stimulation induces a kindling phenomenon that reorganizes a segment of long term memory organization. As we shall see, this phenomenon has implications for pathology formation as well as playing a part in normal processes of system-consciousness change.

To summarize what we have learned so far, we can say that there are seven stages in the process of system-consciousness change:

1. Problem Solving Failure,
2. Error and Novelty Correction,
3. Limbic Triggered Long Term Memory,
4. Fight or Flight Reactions,

5. Kindling,
6. Growth Process and Adaptive Process and
7. New Identity Synthesis in System-Consciousness.

The Intrapsychic Process for Change

The adult has two modes of intrapsychic change, *the growth mode* and *the adaptive mode*. In paradigm terms, the growth mode is a dominant hemisphere process of changing subjective ego functions such as goal formation and planning. The growth mode may also be called *coping*. When there are fundamental changes in the environment or developmental or pathological increases or decreases in subjective capacity, the individual may invoke the *growth process* of identity change. The paradigm reserves the word *process* for modes of change that proceed to the stage of kindling. The dominant hemisphere begins and shapes both the growth mode and the growth process.

The adaptive mode is a means of altering such objective ego-functions as the perception of significance and the regulation of self and object representation. When there are fundamental changes in the environment or developmental or pathological increases or decreases in capacity for objectification, the individual may enter the *adaptive process* of identity change. Here, subprocesses of *internalization* and *identification* determine alterations in subject identity and object identity functions respectively.

Pribram (1981) has written about the adaptive mode of change.

> . . . participatory processes deal with incongruity by searching and sampling the input and accommodating the system to it . . . the experience becomes part of the organism and the plans of actions are appropriately modified . . . Participatory process have in common some kind of involvement, engagement with environment events that extends beyond the organism, but do not operate on them as do motivational processes. (p. 123)

He distinguishes between a predilection for a mode of mental organization that deals with external reality as a *participant* in that reality and another form of identity that contains motivations to act upon reality.

Individual predilection determines the habitual modes of change as well as the processes for change that a person may use in more extreme conditions. The predilection for experiencing life more frequently in dominant or nondominant hemisphere terms appears to be part of the genetic constitution. This has implications for the study of psychopathology. Pribram (1981), for instance, says that overactive participatory engagement, which is here called nondominant engagement, leads to fragmentation of identity, coupled with distractible overinvolvement with reality.

Adaptation and Coping in Relation to Stress

An examination of adaptation and coping provides us with further expansion of the structural paradigm. A great deal of research has been done on the process of stress as a neurophysiological response pattern undertaken when ordinary problem-solving of adult life is ineffective. This research contributes to our paradigm the understanding that a condition of stress occurs when mental representation fails to find or organize salience. In this context, adaptation is therefore the imputation of salience to the new situation. In this sense adaptation is learning to deal with things as they are. Coping, on the other hand, is the willful reconstruction of reality itself so that reality is made to present the desired salient features. Coping is changing the way things are. Adaptation is the nondominant hemisphere's way of responding to stress; coping is the dominant hemisphere's method. The major task of adult life is to use the techniques of adaptation and coping to protect the adult system-consciousness.

As adults, we develop subsidiary intrapsychic identity functions that mediate our participation and purposes in the world. We build, for instance, an affective relation to our own selves as individuals in the community. This relation conditions the sense of *self-esteem* through which we maintain an everyday expectation that people around us will accept our purposes, so that we perceive ourselves as welcome in society or in the family. Self-esteem is thus a guide to the amount of reality drive not currently inhibited by guilt, for guilt is the socially modeled affective inhibitor of the nondominant hemisphere. We develop, too, an indwelling sense of confidence about our expectations of success. This *subjective-esteem* is a measure of our confidence in the availability of a well-tried action-drive for carrying out our purpose. The degree of subjective esteem is a measure of the amount of effort available for the continuation of action plans. Another way of saying this is that the amount of subjective-esteem present is an index of the amount of available action drive that is not currently inhibited by shame. In this context shame is the socially internalized dominant hemisphere inhibitor of expression. The synthetic function of identity maintenance requires, then, that we protect our self-esteem and our subjective-esteem—even at the expense of giving up action plans or particular forms of participation.

There are, however, situations in life, generally stressful ones, in which our usual ways of integrating identity no longer serve to solve the mandatory problems of adult life. Such problems often involve the maintenance of love relationships, the possibilities of doing useful enjoyable work, or have to do with our ability to maintain our standing in the community. When stress (i.e., the failure of problem-solving) occurs, then we enter the process of change we have been talking about.

THE STRESS-SYNDROME

Let us consider the physiological stress syndrome first as a general prototype of changes that occur neurophysiologically during the process of change that supervenes when adult problems cannot be resolved, for the *stress-syndrome* is an overall model for the psychopathological syndromes. Selye (1979) describes this syndrome as having two main stages:

> 1. Alarm Reaction. The organism's reaction when it is suddenly exposed to diverse stimuli to which it is not adapted:
>
> a. Shock phase. The initial and immediate reaction to the noxious agent. Various signs of injury such as tachycardia, loss of muscle tone, decreased temperature, and decreased blood pressure are typical symptoms.
>
> b. Countershock phase. A rebound reaction marked by the mobilization of defensive phase, during which the adrenal cortex is enlarged and the secretion of corticoid hormone is increased.
>
> 2. Stage of Resistance. The organism's full adaptation to the stressor and the consequent improvement or disappearance of symptoms. . . (p. 129)

Although this pattern applies to physical stress, it holds true for mental stress as well, for the same mechanisms come into play. In the countershock phase of mental (as well as physical) stress, the person or animal uses up available catecholamine resources in attempting to find solutions. The emergency use of norepinephrine resources mobilizes an increased output of both ACTH (Adrenocorticotrophic Hormone) and TSH (Thyroid Stimulating Hormone). We may take note of the fact that cortisone, thyroid hormone, and growth hormone are all produced in greater quantities in response to stress, and all have among their effects the mobilization and increase of catecholamines, and increase in receptor sites.

Weiss performed a relevant experiment on mice and dogs that were given, at first, no way to escape an electric shock. In this situation, they became depleted, first of norepinephrine resources, then of dopamine resources. Later, when it became possible for these animals to avoid the noxious stimulus through the experimenter's allowing them new action for escape, the animals failed to mobilize the new action patterns unless they were given two days of rest to renew their dopamine resources. Weiss et al. (1979) concluded from this experiment that

> . . . stressful conditions seem to more readily affect norepinephrine than dopamine, while the mediation of motor behavior seems more closely tied to dopamine. (p. 157)

We may take it from Weiss' work and the work of others that the coping response relates to dopamine and action plan formation. Indeed, we can extend our paradigm further to correlate (1) adaptive response, (2) cortisone, (3) norepinephrine, (4) loss of participatory relevance by the nondominant hemisphere, and (5) amygdala kindling on one hand; and on the other (1) coping, (2) dopamine, (3) growth hormone, (4) failure of action process, and (5) septal nuclei kindling.

Martin et al.'s work (1977) shows that:

> The septal region has a more general role in determining stress responses, not only with respect to ACTH release but also involving GH [Growth Hormone], and behavioral responses. Destruction of the medial and lateral septal nuclei produces hyper-reactivity of both hormonal and behavioral responses to stress stimuli, a finding that led Uhlir and coworkers to suggest that this region is involved in the 'coping' response to stress. (p. 188)

Thus, adaptation and coping instigate homeostatic neurophysiological mechanisms that strengthen system-consciousness.

Understanding Psychopathology through the Neuroscience Paradigm

Various hypotheses have generated experimental findings about psychopathology that are invaluable aids to the understanding of psychopathological syndromes. The following survey looks at these findings as a significant part of the data base on which the paradigm for neuroscience builds. It is meant as a preparation for a full-scale discussion of the left and right hemisphere syndromes.

Before looking at the findings, we must remember that the syndromes must be described both in terms of their triggering processes and in terms of their ongoing development. Take schizophrenia as an example. Evidence indicates that the early, acute pathology resides in dominant hemisphere increase in dopamine processing in temporal, frontal, and prefrontal areas. Later, when chronic schizophrenia supervenes, we find an increased percentage of posterior processing, especially right hemisphere posterior processing. Thus, studies of schizophrenia that fail to specify and distinguish phases of pathology and degrees of chronicity produce inconclusive data on localization. All good localization studies must keep track of the chronology of the syndromes they track.

1. Genetic Hypotheses

In studies of adopted-away children of schizophrenic and depressed parents Kety, Wender, and Rosenthal (1975) tried to show that both depression and schizophrenia originate in genetic predisposition. Using Danish statistics compiled on adopted-away children, they have determined that schizophrenia and depression correlate with heredity. They interviewed relatives about the presence of depression and schizophrenia in the families of origin. They also investigated the presence of allied character types in families of origin and found that the genetic hypothesis of predisposition to one or another type of syndrome held up statistically.

Further refinements of the trait analysis that distinguishes schizotypal from borderline characteristics have allowed researchers Kendler et al. (1981) to align the borderline character with affective disorders and the schizotypal character with schizophrenia. The Kendler group found that when the trait of social avoidance is used to distinguish schizotypal from borderline character, the statistical correlation of schizophrenic tendencies in cohorts of schizophrenics is higher.

These studies refine the original Spitzer and Endicott (1979) trait analysis in these two character types. Still unanswered is the question of what exactly causes the genetic clustering of characteristics that align with these major syndromes. I suggest that hemispheric alignment of functions indicates that the schizotypal character is based on a dominant hemisphere subjective disinclination to form relationships because of the failure to consummate loving relationships, while the borderline character is based on nondominant hemisphere objective failures of the self and object representations to achieve stability sufficient to support social relationships. Hemispheric functional analysis should provide a rationale for further trait distinction.

2. Brain Hypotheses

Scans Anatomically minded researchers have explored statistical correlations between the incidence of schizophrenia and depression on one hand and the incidence of anatomical or physiological variation from the norm on the other. In addition to the examination of the dead specimen in human or animal, methods of anatomical localization now include the CAT scan, PET scan, brain scan, magnetic scan, and the tagged molecule scan. Naturally, such studies require a knowledge of anatomical and physiological norms. For instance, the researchers must know that the right frontal and left temporal areas are usually larger than their hemispheric counterparts (Luchins, 1982). Currently, researchers are developing anatomical maps that localize under specifiable conditions normal physiological brain activity within brain systems. The fact that many norms have not been established makes hypotheses about pathology hazardous. It should also be borne in mind that the information yielded by scanning is only as good as the diagnostic classifications available for correlation.

EEG The relationship of EP (Evoked Potential) responses and CEEG (Computerized Electroencephalogram) responses under specifiable conditions with pathological syndromes also requires a variety of norms. Statistical norms have now identified the sources in the brain of many of the component waves in the EP and have determined the sources of many of the wave frequencies in the EEG. Research that establishes these norms and that uses them must take epigenetic norms into account. The alpha rhythm, for instance, speeds up from 7 or 8 per second at latency to an average of 9 to 12 by adult life (Giannitrappini, 1974). We must also understand that the later components of the EP are frontally determined, whereas earlier components come from a variety of subcortical sites beginning with the cranial nerves themselves.

Finally, to the understanding of anatomical and physiological norms we must add an understanding of *psychophysiological* norms. To give an example, *readiness* and *orientation* are reflex components of the evoked potential expectancy wave. Both readiness and orientation have separate anatomical and physiological meanings as well as separate psychological meaning. The readiness to respond is a product of a dominant hemisphere series of preorganized action-discharge functions. The orientation response is a product of nondominant hemisphere "novelty-set," which is to say of a range of preorganized expectations of reality configurations. When either hemisphere perceives the unexpected, novelty or error, this evokes an expectancy wave of frontal-prefrontal negative potential that readies the system-consciousness to evaluate new information that must be channeled into itself. The wave is evidence of temporary blocking of system-consciousness processing. This prepares the dominant hemisphere to receive new motor sequence possibilities and the nondominant hemisphere to assess a novel configuration of reality. Once the new information has been distributed it is *habituated* and repetitions no longer evoke expectancy waves. This means that the new information no longer desynchronizes frontal functioning.

Psychophysiological definitions are important because the tasks that are measured and mapped are usually understood within a psychophysiological context. A difficulty in evaluating research in this area is that the neurophysiological functions that become relevant in our phenomenology of psychopathological syndromes have not yet been mapped in terms of their hierarchical organization. This, of course, is a reason for the present attempt to develop a paradigm that can include the multitudinous factors in a single framework. Thus, to take as an example research that relies on *conditioning,* we must understand conditioning both in terms of its lower level habituation components and its impact on the higher level ego-functions. If we do not have a context for evaluating research findings, we cannot judge the truthfulness of the hypotheses proferred.

3. Neuropsychological Hypotheses

If the neurophysiologists and the neuroanatomists have been building *up* a map of brain structure systems, the neuropsychologists have been building such a structure *down* to meet them. Since the pioneering work of Sperry, a great deal has been learned about the gross functional systems of consciousness that reside in the hemispheres. The work of Sperry and his pioneering collaborators, Bogen and Gazzaniga (1969) reveals that each hemisphere builds up a separate and disconnected form of consciousness. Each sphere of consciousness thus built can function in an autonomous way, so that in cases of agenesis of the corpus callosum, for instance, individuals can still develop into apparently functioning adults. The surgeon in one of the first cases of hemispheric disconnection (through the severing of the corpus callosum in a case of intractable epilepsy) tells an interesting anecdote that dramatizes this hemispheric autonomy. The surgeon attended a barbecue at the patient's house some years after the operation. While carrying on a polite conversation, the patient picked up a meat cleaver with his left hand, which he brandished threateningly over the neurosurgeon's head. The episode symbolized the disconnection between the patient's affective response to the operation and his intellectual acceptance of the operation. The right hemisphere reenacted an affectively reverberant, traumatic version of the operation while the left hemisphere maintained a well-bound intellectual acceptance (and presumably organized the polite conversation). We must keep this hemispheric autonomy in mind when we consider syndromes in which we believe a single hemisphere is assuming the major coordination function for both hemispheres in contradistinction to the usual synthetic interplay that results in normal behavior.

Major ego-functions such as affect regulation and speech generation can be explored in terms of their lateralization through a number of techniques. In the Wada technique, for instance, sodium amytal is injected into a single internal carotid system, resulting in the disconnection of the affected hemisphere. Neuropsychological lateralization studies of all kinds are hard to assess because it is not possible to know whether the effects are due to facilitation in one hemisphere or to inhibition in the other. Apparently the two hemispheres are mutually coordinated through a process of mutual inhibition. Thus, to give an example, if the nondominant hemisphere is responsible for affect regulation, it carries out its role both through the promotion of nondominant hemisphere processes and through the inhibition or disinhibition of dominant hemisphere processes. Another related difficulty in the search of psychopathological cause in lateralization studies is that without an accurate map of identity functions and ego functions it is hard to say whether one hemisphere is producing a functional result though facilitation of its own systems or through inhibition and disinhibition of functions systems in the other hemisphere. Flor-Henry (1979) presents such a hypothesis in his explanation of psychopathology:

> . . . the fundamental locus of both depressive and manic components of the affective psychoses is situated in nondominant hemisphere systems, but whereas depression is related to nondominant neural structures euphoria and mania is determined by a contralateral loss of neural inhibition—which originates in the non-

> dominant hemisphere—but evokes abnormal activation of the dominant hemisphere through transcallosal projections. . . (p. 682)

I would hypothesize that the nondominant hemisphere regulates and controls, whereas the dominant hemisphere initiates and selects. To this we must add Pribram's contention that the nondominant hemisphere is responsible for participatory processes through which the individual is related to reality by becoming in his own consciousness a part of the process of reality. According to Pribram (1981), an exaggeration of such participatory processes is found in affective illnesses.

Thus, lateralization arguments relate major system-consciousness functions to one hemisphere or the other. Affect regulation and reality participation seem to originate in the nondominant hemisphere. In his summary of the localization of major functional tendencies Flor-Henry (1979) aligns subjective volitional tendencies with the dominant hemisphere.

> . . . subjective consciousness appears to be a function of brain-stem left frontal systems. Papez's conclusion was based on neurosurgical data, and, subsequently some carotid barbiturization evidence has confirmed this relationship. (p. 690)

The major functional dispositions of the two hemispheres provide a context for predicting which other functions belong together.

A neuroscience paradigm is therefore a necessary instrument for providing a context for hypotheses about psychopathology. There is much in the popular literature about the differential hemispheric functions that is misleading because it is sketchy or overly formulaic. It is said, for instance, that the nondominant hemisphere is the creative hemisphere. I believe that this is dead wrong if creativity refers to the initiation and development of new action plans. This popularization has entered into the thinking of psychiatrists without evaluation in the context of a carefully worked out frame of reference.

4. Neurotransmitter Hypotheses

In the last decades, psychiatry has entered a period of neurotransmitter research. The discovery of the major neurotransmitters through tagged fluorescing antibodies has revolutionized our idea of brain systems (Carlson, P., 1959). Very substantial and substantiated hypotheses have linked schizophrenia to defects in dopamine transmission, and affective disorders to defects in norepinephrine transmission. The fact that at least half of the MHPG (the major breakdown product of norepinephrine) found in the urine is due to norepinephrine use and metabolism by the brain, makes this one of the most reliable and nonintrusive gauges of one aspect of brain activity (Maas, 1979). Serotonin metabolism and transmission has been linked to both schizophrenia and the affective disorders in various studies of the change in cerebrospinal fluid levels of 5HIAA (a major breakdown product of serotonin) before and then after acute outbreaks of illness. The fact than any of the three major neurotransmitters has been found to co-regulate receptor protein channels shows the extent to which multiple neurotransmitter control is systematized in the brain.

A variety of hypotheses concerning mental illness have been advanced by students of neurotransmission. A common hypothesis is that a genetic defect causes a disturbance in the formation and/or degradation of the neurotransmitters and so produces false neurotransmitters. Some researchers have looked for particular pathological agents—endogenous brain chemicals capable of acting as false neurotransmitters that precipitate a sequence of pathology. Other researchers search for factors that may initiate supersensitivity in receptor sites.

Other neurotransmitter theorists provide general linkages between the economy of neurotransmitter usage and particular pathological processes by investigating the economy of neurotransmitter metabolism. They have found evidence for statistical deviation in cases of particular pathological syndromes. Breakthroughs in our understanding of mental illness have been limited by the fact that dopamine, norepinephrine, serotonin, acetylcholine, GABA and histamine (another inhibitory system) are all interactive and coactive in many of the same sites. An inhibition in one system will have ramifications in the metabolism of the other neurotransmitter systems through a variety of interacting systems. Specific hypotheses linking syndrome to neurotransmitter metabolism are therefore limited because all of the neurochemical variables cannot be controlled.

Snyder's (1977) exploration of the *endogenous opioid system* adds another dimension to our understanding of neurotransmitter systems. We see that neural tone in various neurotransmitter systems is determined by various hormonal systems, including a widespread endogenous opioid system. This endogenous opioid system shares precursor synthesis with ACTH. Thus, stress affects neurotransmission through a graded series of effects.

The paradigm must include the fact that each of the major neurotransmitter systems has an extended neurophysiological component that is part of its distributed system. The endogenous opioid system, for instance, is a pain-suppressing system that sets the tone in the locus coeruleus. It also fills and blocks many pain receptor sites in the spinal cord, blocking the input of sensory stimulation and pain during acute stress.

Recently, a great deal of work has been done in identifying autoreceptor sites in the major neurotransmitter systems. At issue is the way a cell receives its own neurotransmitter substance. We know that autoreceptors in the nigrostriatal cell bodies of the dopamine system provide a feedback control for the dopamine system (Melzer, 1982). Similarly, alpha 2 receptors in the locus coerulus respond to the presence of norepinephrine by shutting down on the rate of firing of the norepinephrine neurons (Gold, Mark, 1979). Such single-cell feedback mechanisms profoundly effect the economy of neurotransmitters. Because the neurotransmitter cell can both (1) speed up or slow down receptor cells,and (2) speed up or slow down its own firing rate through autotransmission, the amount of neurotransmitter metabolite is not a simple measure of the economies of the neurotransmitter system. This leaves the psychiatrist in the position of not knowing whether tricyclic drugs, say, increase or decrease neurotransmission. Probably they do both. Probably the original effect is for an increase in cortical and limbic transmission and availability. Later feedback alpha 2 receptor blockage must decrease overall norepinephrine metabolism (Kato, 1982).

Another factor to consider is the degree of sensitivity to the neurotransmitter at the receptor site. A degree of hypersensitivity may be induced due to chronic increased neurotransmission in either the norepinephrine or the dopamine systems. Eventually, it appears, the "upregulation" of the norepinephrine or dopamine systems may be interdicted by increased serotonin modulation at the same receptor sites with the induction of receptor site destruction. These complications make any quantitative analysis of neurotransmitter breakdown products doubtful unless the receptor conditions can be specified. Sulser (1984) highlights these points:

> Central β-adrenoceptors are coupled to adenylate cyclase in a stimulatory manner. However there is only circumstantial evidence that alpha 2 adrenal receptors in brain are coupled to adenylate cyclase in an inhibitory manner . . . The occupancy of adrenoceptors by noradrenaline (NA) is a prerequisite for both desensitization of the system and the reduction in the number of β-adrenoceptors while serotonin (5-HT) is co-required with NA

> for the regulation of the density of β-adrenoceptors . . . Since β-adrenoceptor-coupled adenylate cyclase systems function as kinetic amplification systems, small changes in the NA signal transfer are amplified or deamplified respectively. (p. 255)

The energy amplification by the neurotransmitter reception shows the power of these regulatory cell systems.

All of the above factors make the assessment of quantitative neurotransmitter studies difficult. The stage of the illness is also an important factor in assessing the role of neurotransmitters and their facilitation by agonists or inhibition by antagonists. Increased dopamine stimulation, for instance, may exacerbate acute symptoms of schizophrenia such as hallucinations or delusions, while it may lessen the later negative symptoms as chronicity approaches. It appears, further, that D1 receptors in the striatum energize dopamine systems while D2 dopamine receptors may inhibit the energization of target cells possibly through the inhibition of β-adrenoceptor sensitivity.

It seems clear that in depression the β-adrenoceptors become supersensitive, while in schizophrenia the D2 receptors become supersensitive. This means to me that failed reality accommodation in depression and failed consummation planning in schizophrenia lead to increased needs for norepinephrine and dopamine stimulation of the receptors respectively. Once these fronto-active receptor systems have become upregulated in this way, they set a changed overall hierarchical deregulation into motion.

5. Neurophysiological Hypotheses

Neurophysiological hypotheses have been proposed for psychopathological syndromes mainly in connection with a stress model of illness. The reason for this is that considerable animal research was organized around stress assessment followed by the measuring of the various hormonal parameters of stress. As stress has been most closely tied to depression, especially endogenous depression, many of the neurophysiological hypotheses in psychiatry have to do with the role of various hormones in the induction or measurement of syndromatic progression in depression. The focus on the DST (Dexamethasone Suppression Test) as a reliable means of distinguishing depression from other pathological entities is a case in point. The premises upon which Caroll (1981) and others have built and standardized their diagnostic work on the DST is that chronic stress induces a failure of the neuroendocrine axis to suppress the production of endogenous cortisone when a test dose of 1 mg of prednisone is given at 11:00 P.M.

The paradigm helps us to understand the reason for the DST's positive results. The adult system-consciousness maintains intrapsychic as well as neurophysiological homeostasis through successful problem-solving. When problem-solving is chronically unsuccessful, a portion of the frontal-limbic system exhausts neurotransmitter resources and a stress-syndrome supervenes. When the prefrontal regulation fails, the hypothalamic level takes over the regulation of the stress response. As we have seen in the case of deepening endogenous depression, hyperproduction of cortisone has the homeostatic effect of increasing neurotransmitter production. This tends to restore the competency of system-consciousness. Other hormones are also deregulated from prefrontal control in stress. Thus Cowdry (1983) finds that the thyroid stimulating hormone is decreased in bipolar patients. Like cortisone, thyroid hormone is a neuromodulator. Thyroid hormone acts to increase beta activity in the norepinephrine system and to decrease alpha 2 activity. This means that the increase in thyroid hormone early in the stress response increases the available expendable norepinephrine, whereas chronic cycling depression results in a hypothyroidism that causes an increase in alpha 2 activity which leads to an overall decrease in the activity of the norepinephrine system. This causes a conservation of resources in prolonged stress.

Snyder (1984) reports that endogenous opioids are highly concentrated in the amygdala and in the locus coeruleus. Opioids are stress related comodulators acting along with other neurotransmitters in the regulation of brain system activity. The fact that opioids are produced along with the cortisol releasing factor as a sequence of amino acids on a precursor protein means that in stress, neuromodulators are produced that tend to channel and direct the stress response. In the case of endogenous opioids acting at the amygdala and the locus coeruleus we see that the inhibitory effect of these opioids tends to shut down system-consciousness processing in favor of more instinctual responses. The concentration of these opioids in the norepinephrine system regulating centers indicates once again that the pain system is an extended neurophysiological component of the norepinephrine system of regulation. The effect of neuromodulators must be considered in any quantitative appraisal of neurotransmitter system functioning.

High level cortical and thalamic regulation of hypothalamic and lower level neurophysiological functions are mediated by many neuromodulator hormones. As serotonin is exhausted somatostatin follows suit, releasing growth hormone (Rubinow, 1983). Growth hormone and prolactin both appear to have evolved a special feedback regulation of dominant hemisphere processes. Action and coping distress produce an increase in both of these neuromodulators. Prolactin may be, in my opinion, a messenger hormone, for it is released in greater quantities whenever upper level cortical control is decreased. Prolactin is increased when the cortex is damaged. Cortisone and thyroid systems of hormonal modulation, on the other hand, appear to have evolved a feedback regulation of nondominant hemisphere processes. Increased stress produces an increase in these hormones with a feedback increase in norepinephrine resources to process reality.

The amount of all of these neuromodulators, their level of regulation, and their response to the diurnal cycle must be related to syndrome development from acuteness to chronicity.

Dominant Hemisphere Processes

Schizophrenia

Let us take schizophrenia as the model for dominant hemisphere-triggered processes of pathology. Our hypothesis is that schizophrenia represents a long term inability to cope with the age-appropriate task of falling in love and expressing love in a subjectively satisfying way. As an illness that usually begins in middle or late adolescence or in early adult life, schizophrenia is a growth process of identity change that goes awry.

The dominant hemisphere is responsbile for dopamine processes of regulation of consummation strivings. When such strivings are blocked, love cannot develop into an ongoing process, so that the overall subjective function of identity regulation cannot mature. The preschizophrenic is therefore beset by identity struggles, failures to love, humiliation, shame, and feelings of loss of subjective integrity of self. A claustrophobic anxiety-panic of blocked action plans and immobilized subjective life continually increases. During this prodromal stage, dopamine resources build up and are stretched to the limit as D2 receptor sites become supersensitized (Crease, 1983). The acute panic stage exhausts serotonin and gabaminergic resources for inhibition of the action drive. Subjective identity deneutralizes them, so that psychosis follows.

Consider some evidence for this paradigm view of acute schizophrenia. Kornhuber finds that several features of the EEG correspond to the schizophrenic's difficulty in initiating voluntary movement (Kornhuber, 1983). When Abrams and Taylor (1979) compared cases of affective psychosis and schizophrenia in EEG studies they found a significantly higher proportion of abnormal EEGs, distinguished mainly by slower waves in the left temporal lobes of schizophrenics. Their group of schizophrenics were subacute, showing major Schneiderian signs of schizophrenia. This suggests that the left septal amygdalar area was previously highly kindled in acute schizophrenia; but that in the subacute stage the kindling is spent which produces some diffuse slowing. Using a computer analysis of the spectral power of the EEG, Gianitrappini and Kayton (1974) analyzed the components of the CEEG in subacute young schizophrenic patients. They found a lower average rate of alpha activity in the 8 to 10 cycle activity range, compared with a normal activity of 10 to 12 per second. The authors took this as presumptive evidence for a reversion to earlier modes of processing, since alpha rhythm activity tends to speed up with maturation. According to them,

> alpha activity furnishes a rhythm for a generic search for stimuli which ceases or subsides in the presence of a broad range of percepts, whether internal or external. (p. 383)

They conclude that the schizophrenic gives up on the search for satisfiers. This interpretation complements the paradigm understanding that the mental apparatus functions to match goal-oriented consummation strivings with specific features of the satisfying object. Presumably the subacute stage develops as the schizophrenic develops organic and psychological defenses against the humiliation and rage of failed consummation.

There is substantial evidence for the relevance of changes in dopamine metabolism in the genesis and unfolding of the schizophrenic syndrome. Kato and Ban (1982) have summarized the evidence for the dopamine hypothesis of schizophrenia. First, the efficacy of neuroleptic drugs in schizophrenia is related more to the drugs' dopamine blockading activity than it is to any of the neuroleptics' other functions. Second, there is an increase in the number of dopamine receptors in the limbic areas even of schizophrenics who have never been given medication. This indicates that the dopamine system for promoting consummation is overdeveloped in the prelude and early stages of schizophrenia. Post et al. (1975) have found that the amount of the major dopamine metabolite HVA (Homovanillic Acid) is decreased in the spinal fluid after the acute stage of schizophrenia has passed. The Post group also found that in the subacute stage of schizophrenia, both the dopamine metabolite and the serotonin metabolite 5HIAA decrease markedly in the spinal fluid. This information supports the hypothesis that the acute stage of schizophrenia exhausts the serotonin capacity. The ability to neutralize is therefore diminished and thus identity functions, no longer integrated, become unable to integrate the ego-functions. As dopamine transmission then decreases in the subacute stage, we see deregulation of subjective identity functions, and a loss of regulation of speech generating process that leads to the *defect in associations* found in schizophrenic phenomenology.

Thus, we may view the movement toward chronicity in schizophrenia as consisting of a gradual shutting down of dominant hemisphere consummation strivings. Using computerized tomography in chronic schizophrenic patients Luchins found that schizophrenics have a thickened corpus callosum, and that they fail to show the normal asymmetry of right frontal enlargement compared to the left frontal area (Luchins et al., 1982). Conceivably manifesting a hypertrophy of defensive inhibition this helps explain the eventual blockade of left prefrontal functioning in schizophrenia.

Using measurements of regional cerebral blood flow in comparisons of acute with chronic schizophrenics Ingvar (1974) has found, and Ariel (1983) has confirmed, that the chronic schizophrenic shows a poverty of frontal activation and a shift to right posterior processing, features not found in acute schizophrenics. Ingvar tied his findings to the chronic schizophrenic's hypointentionality: ". . . the lower the flow was in premotor and frontal regions, the more striking were symptoms of indifference, inactivity and autism." (Ingvar, 1974, p. 1485). Thus, Ingvar's conclusions support the contention that the loss of dominant hemisphere identity functions and ego functions leads to the chronic state of subjective impoverishment, which includes lack of goal orientation, planning and initiative. Anyone who works clinically with chronic schizophrenics can confirm that this hypointentionality is the major obstacle to therapeutic success.

Dominant Hemisphere Character Distortion

An understanding of dominant hemisphere character typology follows from consideration of the dominant hemisphere functions. The avoidant personality avoids social contact because it is humiliating. The schizoid character avoids social contact because it has become unconsciously dangerous. The narcissistic character uses social contact only to attempt to strengthen the subjective sense of grandiosity and power. Finally, the paranoid character shuns social contact because of a sensitivity to criticism and a readiness to feel subjectively shattered that exceeds even the narcissistic character's. Thus, as is not noticed in *DSM III*, one whole spectrum of character pathology centers on failure to consummate love strivings. Using the paradigm we can organize our thoughts about dominant hemisphere character types around the concept of the dominant hemisphere identity functions. In dominant hemisphere pathology, the identity functions are distorted in the direction of (1) a failure to consummate love, (2) involvement with rage and hatred that follows in the failure to consummate love strivings, and (3) fixation on earlier life modes of attempting to consummate expressive and love strivings. The grandiosity seen in these character distortions must therefore be seen as an attempt to replace present failures in consummation with past ideals and wishes for consummation, with earlier dominant hemisphere modes of relating.

In this context, we may take the schizotypal character as one generalized dominant hemisphere character. In *The Search for a Schizotype* Gunderson et al. (1983) re-examined the records of the Danish studies of extended families of twins to find a better cluster analysis to define schizotypal characteristics. In this way, they hoped to align the schizotypal character as a prototype character that is genetically correlatable with schizophrenia. They found this correlation to be true if the traits of social withdrawal and avoidant behavior are added to the schizotypal character, and if the traits of magical thinking and dissociative episodes are subtracted. Thus, the schizotypal traits of (1) social isolation, (2) anxiety, (3) suspicion, and (4) odd behavior can all be taken as related to the failure to maintain stable subjective identity, and to the failure to fulfill the dominant hemisphere ideal of loving another person.

Nondominant Hemisphere Processes

In the nondominant hemisphere model of identity regulation Pribram describes, the individual easily enters into the process of identifying with and internalizing the characteristics of other people and of the surrounding society. These television watchers, consumer responders (in contradistinction to dominant hemisphere personalities), are field dependent types who react adaptively to changes in their environment by trying to recreate these changes in themselves.

Dominant hemisphere personalities are more field independent, more concerned with their own subjective methods of coping, and more prone to attempt a creative solution through changing their whole subjective approach to life when they are stressed. Nondominant hemisphere persons feel that others hold the key to satisfaction.

When the nondominant hemisphere type of person is stressed he or she tries (1) to take in ever more data concerning reality, (2) to organize this data into salient features, and (3) to use these features in a search for the ending of frustration. Nondominant hemisphere characters tend to react affectively and to use their affective reactions to convince others that they are deserving. They also use these reactions as self signals about the kind of changes in image and representation of self required to find gratification. When frustration persists, their tendency is to become anxious in the agoraphobic way, to feel abandoned. They fear they will be deserted and that there will never be another person to mediate their relationship to reality. Sociobiological studies of life events disposing to depression emphasize these factors. Paykel (1979) shows that "exits" of other persons are the precipitants of affective disorders, while Brown (1979) shows that (1) the loss of a mother before the age of eleven, (2) child rearing by a mother without a mate, and (3) general social isolation are the disposing factors for major depression.

The failure of agoraphobic anxiety and affective response to stem social and personal abandonment leads to the acute stage of major depression. In this stage the individual exhausts serotonin resources available to maintain an authentic feeling identity as part of the social surroundings. The increased norepinephrine utilization in the search for salients supersensitizes the central beta receptors, while the co-modulation of serotonin with noradrenaline at these sites is overcome, leading to a failure to 'neutralize' nondominant hemisphere identity functions. The failure to maintain objective identity function leads to the acute symptomatology of major depression. As the system-consciousness fails in its regulatory function, guilt increases pathologically, self-esteem lowers pathologically, and feelings of hopelessness rise. Sad affect, agitation, or retardation (slow movement) may develop, all of which depends on the amount and distribution of norepinephrine and serotonin resources available. The loss of system-consciousness regulation and the exhaustion of serotonin and norepinephrine leads to sleeplessness, premature REM, and finally to psychotic decompensation, accompanied by simple identity-related auditory hallucination consisting of having one's name called. The lost object may persist beyond dream imagery into visual hallucination during the day.

During this subacute phase of psychotic depression, the DST becomes positive as cortisone and thyroid output increase. During the subacute stage cortisone and thyroid enhance norepinephrine production. TRH and ACTH and other brain neuropeptide modulators also enhance norepinephrine activity for a while. However, as depressives move toward chronicity and hormonal resources are depleted, the patient experiences a hopeless feeling. Affective responsivity diminishes, and the individual goes into a slower metabolic state, more or less to wait out the traumatic situation until such a time as hope and enhanced neuromodulator and neurotransmitter resources may return.

This description is consistent with the paradigm view of the nondominant hemisphere's functioning. Consider some evidence for this view. Nondominant hemisphere ECT (Electroconvulsive Therapy) is effective in the alleviation of depression, whereas unilateral dominant hemisphere ECT is less so. Flor-Henry (1979) conducted a research study that compares CEEG in schizophrenics, schizoaffectives and manic-depressives. In it he found that "both mania and depression were characerized by asymmetrical frontotemporal dysfunction; implicating the nondominant hemisphere" (p. 683). Flor-Henry also found that the depressives had irregularities in the higher frequency power spectra. These were quite marked in the right temporal area and were also present, though less markedly, in the left temporal area. This suggests, within our paradigm, that the painful affects of depression are particularly engendered by right amygdala kindling. This hypothesis is consistent with the theory that depression is a syndrome of repeated trauma-induced kindling that produces forced input of stimuli and features from reality that is experienced as forced internalization of sensory imagery. Certainly, this entire view is supported by practitioners' experience with traumatized patients who relive all the perceptual stimuli involved with the traumatic situation until it is all finally bound and included within ordinary mental process. Kindling enhances neuromodulator facilitation of new memory.

In a third line of evidence for the implication of the nondominant hemisphere in triggering depression, Flor-Henry shows that depressives exhibit a higher power spectrum in the right parietal area that we can interpret as a forced internalization of spatial cognitive data, for the right parietal area functions to process visual spatial data from reality.

This evidence of increased emphasis on posterior right hemisphere parietal and temporal lobe processing dovetails with a slightly different line of experimental inquiry. Shagass et al. (1978) find in evoked potential studies with psychotic depressives, schizophrenics, and nonpsychotic patients, that psychosis in general is associated with a decreased amplitude in the later components of the evoked potential response. If we take the later components of the evoked potential as generated by the frontal cortical areas, then this is evidence for a reduced frontal processing of information during the more chronic phases of psychotic processes when identity functions have been compromised. There was no evidence of reduced amplitude in the evoked potential waves in nonpsychotic patients with minor depressive disorders. Thus, Shagass' data re-enforces the view that all chronic psychosis, wherever triggered, results in decreased overall frontal processing and a shift to right hemisphere processing.

The norepinephrine hypothesis of affective disorders has led to a great deal of data acquisition that relates an impoverishment of norepinephrine to the onset of major depression. Bipolar patients have been found to have lower overall levels of norepinephrine and more fluctuation in the amount of available norepinephrine before and after an attack of mania. Thus, Schildkraut (1978) finds lower MHPG urinary excretion in bipolar patients than in unipolar depressed patients. Endogenously depressed patients have low MHPG levels. If we take endogenousness as a measure of syndrome chronicity, then we may take this information as an indication that norepinephrine is depleted in the long term overzealous searching of reality. Certainly bipolar patients are notorious for their overinvolvement with reality perception during the acute stages of their illness.

Recent research has centered on the role of the locus coeruleus in setting up the conditions for depression. Saric (1982) emphasizes that the norepinephrine system in the locus coeruleus mediates the relation to the external world. Thus, overstimulation of this center may be taken as one of the trigger points in the depressive process. In chronic stress natural opioids, acting as cotransmitter at the locus coeruleus, inhibit the prolonged energization of this central acting noradrenergic system. It has been found that the therapeutic effect of the tricyclic drugs is connected with their inhibition of the alpha 2 receptors of the locus coeruleus. This cuts down on the activity of this neurotransmitter system. When tricyclics are withdrawn, patients experience a rebound phenomenon with increased activity of the locus coeruleus and an increased level of norepinephrine activity. Thus, Charney et al. (1982) have found that stopping tricyclics produces a higher MHPG level and increased anxiety. We see in this rebound, then, a return to the high level of novelty-anxiety that occurred in the prelude to the outbreak of the major depression. Sometimes termination of tricyclic drugs even leads to manic episodes. This re-enforces Bunney's conclusions concerning *The Switch Process in Manic Depressive Illness* (1972). He found an increase in both dopamine and norepinephrine metabolism, accompanied by a decrease in serotonin metabolism—which amounts to a dramatic metabolic switch—at the onset of mania. According to Sulser (1984)

> . . . the demonstrated biomolecular linkage between serotonergic and noradrenergic neuronal systems in brain at the level of the NA receptor-coupled adenylate cyclase provides a scientific basis for unifying the two main hypotheses of affective disorders into a "serotonin-noradrenaline link hypothesis" of affective disorders. (p. 257)

The dramatic mania onset is one of those clinical-organic processes that convince us that psychological process is brain process. The switch is accompanied by massive psychological denial of reality. It is sometimes accompanied by a blank dream and orgasm. Orgasm usually involves a nondominant hemisphere amygdala discharge (Flor-Henry, 1979). Thus, the manic switch indicates a massive nondominant release of inhibition, facilitating dominant hemisphere processes. The increase of dopamine metabolism in the acute stage of mania re-enforces our paradigmatic expectation that the hypersexuality in this illness is due to a dominant hemisphere response to repeated release from the discharge inhibition usually maintained by the nondominant hemisphere. We have not so far considered the impoverishment of the serotonin system that occurs in the acute stages of both psychotic depression and mania. The fact that some drugs for the treatment of depression stabilize the serotonin system while others stabilize the norepinephrine system indicates that both of these systems are involved in the bipolar process. Brown et al. (1982) report that the more violent and aggressive patients in their studies showed lower 5HIAA spinal fluid levels. They also find that the more action prone patients tended to show higher levels of MHPG. This data indicates that serotonin is involved in the neutralization of behavior and in the stabilization of identity functions. We may take the depletion of serotonin accompanied by the increase in catecholamines in violence and mania as a confirmation of the view that serotonin neutralizes the drives and stabilizes the system-consciousness. This neutralization is undoubtedly related, in part, to co-regulation at adenylate cyclase receptor sites where central noradrenergic influences are facilitatory. Thus, lithium may block mania through the increased serotonin synthesis that follows its use. Chouinard et al. (1983) point out that clonazepam, which also increases serotonin levels, blocks manic symptoms. Meltzer et al. (1984) found that affective patients responded to a dose of serotonin precursor (5HTP) with increased cortisol production, indicating, in his opinion, that serotonin receptors are supersensitive due to their depletion in the acute affective disorders.

NONDOMINANT HEMISPHERE CHARACTER PATHOLOGY

One of the findings of modern psychiatry is that a variety of apparently unrelated syndromes have now been found to have nondominant hemisphere aetiology in common with the affective syndromes. Thus, bulimia and anorexia (Pope, 1983a), obsessive compulsive syndrome (Insel, 1983), and the dysthymic disorders (Akiskal, 1983), have all been found to respond to the tricyclic drugs. It has also been found that genetic relatives of patients having these syndromes show statistically higher prevalence of major affective disorders. Even more telling is the fact that the borderline personality, as it is defined in *DSM III*, is highly related to affective disorder. In fact, Pope et al. (1983b) have found in a study of borderline patients over a period of years that more than fifty percent develop a major affective disorder. What is more, Soloff and Millevard (1983) found a significantly high prevalence of affective disorder among the relatives of borderlines. Thus, we know that borderline personality breaks down into affective psychosis, rather than into schizophrenia or schizophreniform illness.

The term *borderline syndrome* now refers to a comprehensive syndrome that includes a cluster of personality traits and predispositions that relate to nondominant hemisphere functioning. Some hysterical characteristics, many obsessive ones, many dependent characteristics, most masochistic characteristics, and many sociopathic characteristics are clearly related to the clustering of traits found in the borderline syndrome. To be sure, the hysterical personality, like the paranoid personality, shows mixed hemispheric distortions. Nevertheless, a consideration of the identity functions, the ego functions, the neuropsychological functions, and the neurophysiological functions organized by the nondominant hemisphere prefrontal areas will give us additional insight into the nature of syndromes that are nondominant hemisphere related.

In addition to a genetic disposition for nondominant hemisphere participatory processing, we must include a psychological cause and a developmental cause in a description of borderline development. In their study of the characteristics of borderline families, Gunderson et al. (1980) find that the parents of borderline children are preoccupied with one another to the exclusion of their offspring. The children therefore lack the mediation of their reality, and attention to what they needed to learn about reality, to grow up successfully and to know how to adapt to changing conditions. These families also tended to show overt maternal hostility toward the children. Under these conditions, the borderline child must develop a 'bad' self image, and split the parents into an actual bad parent and an idealized good parent, for in the absence of 'good enough parenting' every child will overdevelop an idealized good parent.

The psychoanalytic hypothesis that relates to nondominant hemisphere processes follows Freud's conclusions in *Mourning and Melancholia*, (Freud, S.E. Vol XIV), which, in its conception, is a thoroughly modern paper. As Freud showed, the loss of the ambivalently held, or split off, object leads to a process of pathological identification. If we understand that such processes are primarily fostered by the nondominant hemisphere, as a response to traumatic conditions, then we may see borderline development as chronically traumatized development leading to distorted characterological responses to reality. The suicidal, self-destructive tendencies in borderlines may be seen as the result of angry affective responses directed against self-structure that is pathologically identified with a whole developmental series of lost traumatizing objects.

In their paper describing the development of criteria for the borderline diagnosis Spitzer and Endicott (1979) emphasize the unstable affect and the unstable identity-sense in borderline patients.

> . . . Schizotypal personality was chosen to represent the concept of borderline schizophrenia, since the term means "like schizophrenia." We chose the term "unstable personality" to represent the concept of borderline personality, since most descriptions of such patients referred to unstable affect, interpersonal relationships, job functioning, and sense of identity. (p. 18)

The unstable hierarchy of functions in the borderline's nondominant hemisphere gives rise to the symptom formation. Indeed character organization itself may be seen as a pre-eminently and globally nondominant hemisphere ego function.

In their study of the obsessive-compulsive disorder, Insel et al.

(1983) find that the drug clomipramine increases the availability of serotonin, and so improves obsessive-compulsive symptoms, whether or not depressive symptomatology was overtly present. Such findings lead to the hypothesis that obsessive-compulsive illness occurs when nondominant hemisphere neutralization is decreased. A consideration of nondominant hemisphere neuropsychological functions makes this hypothesis clearer, for these functions include (1) the organization of perceptual data concerning salient signifiers of satisfaction and (2) a decision-making process about the possibility of completing action-plans. These are precisely the functions that give rise to the clinical symptomatology in obsessive-compulsive illness. Indeed, it goes without saying that the obsessive-compulsive person is overly concerned with trivial details in reality. The obsessive cannot decide which details are significant, and therefore cannot make a decision about action-plans. This person solicits advice from everyone in the attempt to resolve doubts about possible courses of action. This distortion of character in the obsessive-compulsive syndrome leads to a hypertrophy of nondominant hemisphere object-identity functions, that in turn leads to the overtaxing of the norepinephrine system for binding perceptual details, and the serotonin system for bringing such details into consonance with action-plans. Tricyclics can block the overactive norepinephrine-forced overbinding of reality.

A Theory of Psychosis

Our paradigm gives rise to theoretical formulations concerning pathological processes that combine the organic and psychological contexts. The advent of *DSM III* has made us much more aware that all forms of psychosis; schizophrenic, schizophreniform, depressive, manic, and indeed organic, can manifest the same major symptoms. Thus, we must develop a theory that indicates why the brain responds both to overwhelming psychic conflict and to organic disruptions of high level functions with the formulation of primary delusions and hallucinations. Janice Stevens (1982) has moved in the direction of an explanation through her obervations of the inherent opposition between psychosis and epilepsy. She points out that in temporal lobe epilepsy, the tendency to psychosis is greater when seizures occur less frequently, and the obverse, that when seizures are frequent the tendency to psychosis is less:

> . . . the catecholamine systems exert a powerful inhibitory effect on propagation of EEG spike activity and may thus limit excitation to specific regions of the brain in which the spike mode of signal transmission is physiologic . . . "Kindling" . . . is associated with progressive changes of behavior prior to development of a chronic epileptic state. (p. 260)

It is likely that decreased drive system control leads to either behavioral abnormality or psychosis. The neuroleptic blockage of behavioral breakthrough increases the likelihood of seizure activity.

The fact that catecholamine blockers used in the treatment of schizophrenia lower the seizure threshold is one indication that psychosis is a compromise between extreme and dangerous behavioral discharges on one hand, and unrestrained neurological system activity arising in the limbic system that leads to seizure discharge and autonomic deregulation on the other. We may surmise that uncontrolled conflict produces an increase in limbic kindling that triggers a structural change in the system-consciousness, so that the unregulated instinctual forces cannot be inhibited. The change in system-consciousness that is produced as a compromise protection against unrestrained dangerous behavior or seizure activity is in the direction of increasingly generalized synthesis at the expense of the integration that occurs within the individual zones of the system-consciousness. By this I mean that the separate identity functions of subjectivity, agency, self, and object integration merge, producing primary delusions with the inability to locate thoughts within the integrated identity structures. In the nondominant hemisphere the merging of self and object functions produces a defect in reality testing and processing, as well as depersonalization and derealization. The merging of identity functions in the dominant hemisphere produces a feeling of expanding or contracting subjective identity (grandiosity) and an inability to form ordinary goals. Thus, hierarchical overgeneralizing identity synthesis accounts for many of the clinical phenomena in acute psychosis.

Further symptom formation in psychosis is caused by the breakdown and lack of integration and regulation of the ego functions that are subservient to the identity functions. Thus, the generating speech system that is organized by the identity function of agency as volitional planning loses coherence giving rise to the schizophrenic type of hallucinations. In early schizophrenic hallucinations, it seems that the individual hallucinates action plans that are separate from the volitional awareness of subjective generation.

The breakdown in identity synthesis is manifest in most case histories. Consider the case of a schizophrenic patient who threw a brick through a window, an event that lead to his hospitalization. The patient declared that he threw the brick through the window to avoid throwing the brick at a policeman whom he believed to be his father. Three years earlier the patient had found himself unable to repress a wish to kill his father, who happened to be a policeman. To avoid his wish the patient went through a psychotic process that led to a generalization of his father, a generalization of his self-image, a delusional belief that his mind could be read, and a delusional belief that there was a 'mafia organization' that controlled his mind. He represented this organization, and the structure of his whole psychosis as a "ring". Under sodium amytal he told us that the ring was a mafia ring that included the police. It was the ring that weds male and female (which were united in a hermaphroditic image in the patient's mind), it was the ring of his thoughts going around and around, producing the sound of his thoughts themselves.

Typically, as in the case vignette, the structure of every psychosis contains an intrapsychic description of the psychotic process itself. To give a further example, the belief common to psychotics that God organizes and influences the patient's individual mind comes from a defensive compromise formation. The formation involves an overgeneralization of all identity functions, which defends the patient against behavioral breakdowns that may result in manifestations of decontrol of primitive conflicts and fantasies. Thus, insofar as they are utilitarian, the psychoses protect from murder, suicide, rape, homosexual activity, incest—and the whole gamut of lifelong fantasy formation that requires a prefrontal system-consciousness for its regulation and secondary process for its organization.

The theory that psychosis advances rapidly toward more generalized synthesis of identity functions that travesty the appropriate formations for middle and old age rests upon the paradigm belief that as the process of human epigenesis goes forward there is a hierarchical development of the prefrontal cortex that produces higher, and more generalized levels of regulation of consciousness during each period of life. This overrapid progress takes place when the identity functions of system-consciousness fail to provide problem-solving in situations of insoluble conflict. Under these circumstances, a person must try to use all the available organic resources of hierarchical generalization. When these resources fail to hold, the psychotic process takes a regressive course that emphasizes either dominant hemisphere growth process or nondominant hemisphere processes of adaptation.

Presumably, during the regressive course of the psychotic process, intercurrent kindling occurs—more left kindling in acute schizophrenics, more right kindling in acute affective psychosis. As Stevens (1979) points out,

> in all mammals examined, the amygdala and the hippocampus have the lowest threshold for seizure discharge of any region of the brain. (p. 260)

She adds:

> If, as the evidence suggests, catecholamine systems normally restrict physiological hypersynchrony and spike activity elicited by specific neurotransmitters to specific brain areas, the physiological increase in excitability of amygdala circuits accompanying sexual maturation or stress must be accompanied by increased activity in local inhibitory systems to limit subcortical excitation to appropriate limbic and axial structures. (p. 260)

Stevens' statement relates to the circuitry of the interaction between the limbic triggered discharge systems and their prefrontal cortical control by system-consciousness apparatus for prefrontal regulation that includes the norepinephrine system, dopamine system, serotonin system, and the GABA systems. When the prefrontal system-consciousness fails to regulate the limbic circuitry, then kindling occurs with the intake of new experience—"significance experience" in the case of schizophrenia, and forced intake of reality in the case of the affective psychoses. This stage of the psychotic process, which corresponds to the subacute stage, fixes a psychotic identity structure. The formation of psychotic identity structures furthers the psychotic process. Thereafter, drive processes are increasingly defended through relinquishment of the drive demands on system-consciousness. Chronicity is a state of extreme identity generalization and economy that reduces the pressure for supersensitive catecholamine drive reception, and thus allows the re-establishment of neurophysiological homeostasis.

Thus, in the chronic stages of either form of psychosis, frontal processing decreases, and intrapsychic life is maintained on the basis of poorly differentiated highly generalized structures of identity that regulate the posterior life-sustaining neuropsychological functions. The overall metabolism of brain-functioning decreases in the chronic stages of psychosis, much as it does later in human epigenesis, when neurotransmitter energy is less available because of catecholamine system cell death and degeneration. This means that the chronic stages of psychosis are a process of using later life stage mechanisms to conserve mental life. Shagass et al. (1978) find that the later components of the evoked potential are reduced in all forms of chronic psychosis. Prefrontal and frontal functioning, setting the mind to solve problems, is de-emphasized in chronic psychosis.

Conclusion

I would like to conclude this description of the structural paradigm for neuroscience with some remarks about the anxiety systems and their relationship to the stabilization of system-consciousness. The preservation of system-consciousness and its attendant intrapsychic identity synthesis is a matter of life and death. The two anxiety systems function to inhibit drive pressures that threaten the maintenance of this system. Thus, in every conscious manifestation of anxiety, we see either an anxiety that subjective life will fail, or an anxiety that the participation in reality will fail. The dominant hemisphere forms of anxiety—error, anxiety, and claustrophobic anxiety—are subjective anxieties that inform consciousness about the state of preservation of the integrity of subjective life and the synthesis of mental life. Similarly, the nondominant hemisphere forms of anxiety—novelty anxiety, stranger anxiety—agoraphobic anxiety—indicate that the objective foundations of continuing life are threatened. Consciousness of anxiety occurs in all forms of psychosis and in all of the anxiety disorders. Increased understanding of the hemispheric nature of mental organization will allow us to understand the nature of the communication inherent in phenomenology better and thus increase the sensitivity of our differential diagnosis. Such an understanding will help us understand the pathology of brain process in lieu of our classical diagnostic terminology. Finally, I would like to present a chart of brain/mind congruence that shows the functional hierarchy as it is lateralized and that shows the relationship of lateralized functions to particular syndromes.

Table 2
Brain/Mind Functional Map (System-Consciousness)

Structure	*Prefrontal Dominant Hemisphere*		*Interhemispheric*	*Prefrontal Nondominant Hemisphere*	
Architectonic Zone	9, 45, 46 (lateral)	9, 10, 11 (medial)		9,10,11 (medial)	9, 45, 46 (lateral)
Energy-System	Dopamine (Action-Drive)		Serotonin (Neutralizing-Drive)	Norepinephrine (Reality-Drive)	
Regulatory-System	GABA (Action-Anxiety)		Acetylcholine (Facilatory-Gating)	GABA (Reality-Anxiety)	
Function	*Action Program Integration*		*Program Synthesis*	*Reality Program Integration*	
Identity-Function	Agent-Integration	Subject-Integration	Mental-Synthesis	Self-Integration	Object-Integration
Ego-Function	Volition	Expression	Interpersonal	Affect	Perception
Organizes	Cognition	Emotion	Judgment	Emotion	Cognition
Integrates	Speech	Consummation	Communication	Feelings	Character
Neuropsych.-Function	Performance	Pleasure-signal	Attention	Pain-signal	Representation
Organizes	Meaning	Wishing	Salience	Need	Familiarity
Regulates	Sequencing	Effort	Selection	Concentration	Grouping
Memory-Function	Event-formation	Fantasy-formation	Program-suspension	Affect-reverberation	Episode-formation
Organizes	Learning	Conditioning	Habituation	Conditioning	Learning
Regulates	Future-Time	Past-future-time	Time-Selection	Past-present-time	Present-time
Neurophys.-Function	Sympathetic		Parasympathetic	Sympathetic	
Organizes	Readiness-response	Fight-response	Di-urnal-Rhythm	Flight-response	Orientation-response
Regulates	Prolactin	GH (Growth Hormone)	Somatostatin	ACTH	TSH
Development	*Creative Process (Growth)*		*Epigenesis*	*Adaptive Process*	
Subprocesses	Coping	Regression	Primary Process	Progression	Identification
Pathology	of Error-Detection	of Love-Consummation	of Problem-Solving	of Frustration-Tol.	of Novelty-Detection
Psychosis	Schizophrenia		Organic	Affective	
Anxiety-Syndrome	Claustrophobic			Agoraphobic	
Character-Disorder					
Fixated	Narcissistic	Hysterical		Masochistic	Obsessive-compulsive
Traumatized	Avoidant	Schizotypal		Borderline	Passive-aggressive
Deprived	Schizoid	Paranoid		Dependent	Antisocial

Chapter 2

A BRAIN-MIND FRAMEWORK FOR PSYCHIATRIC NOSOLOGY

This chapter offers a brain-oriented framework for psychiatric nosology (i.e. classification of diagnoses). It uses the paradigm to redefine psychiatric syndromes, both in terms of their hemispheric and zonal triggering and in terms of their localization tendencies during the disease process. It is a deliberate attempt to replace the present system of diagnosis, which is based on presenting symptoms rather than cause. The present nosology finds cause by matching symptomatology with the paradigm model for hemispheric and zonal function and dysfunction and for distortions in the basic mind-brain processes, which the paradigm defines as processes for growth and adaptation. This nosology tries, in other words, to cut through the problem of shifting emphasis and styles of diagnosis. DSM III, for example, is now extinguishing an historically recent tendency to diagnose most psychosis as schizophrenia, but it is also de-emphasizing the meaning and presence of schizophrenic illness. The present chapter's essential work is to find an order, based on cause, in the present diagnostic system that DSM III represents. It is hoped that this reordering will give researchers a map of general areas of brain/mind coordination that will assist increasingly precise research.

THE FUNCTIONAL HIERARCHY

To understand the clinical phenomena under investigation, we must review the hierarchy of functions that the paradigm outlines.

The Identity Functions

Making up the hierarchically uppermost level of the coordinated system-consciousness, the identity functions comprise the four integrative functions that form the basis of identity synthesis. Their two tasks are (1) the regulation of the discrete ego-functions located in the distributed system of each identity zone and (2) the coordination of the zones with each other. This is to say that the identity-functions are the means by which the mind signals the brain. While the identity-functions receive all brain signals, they themselves inaugurate fresh brain processes. They are the sole inaugurators. Their function is to further preserve the individual's existence. Working together, the four zones create the reflective sense of personal awareness through which we recognize ourselves as (1) active, (2) subjective, (3) responsive, (4) distinct human beings in the world.

(1) The Agent Function This dominant hemisphere lateral identity function maintains action-planning and goal-orientation. It engenders the sense of initiative that is fundamental to pragmatic action. The agent is the source of volition. The meaning inherent in motor sequence gives rise to the subjective sense of the body as a tool. The agent controls the use of this tool, and its extensions—real tools.

(2) The Subject Function This dominant hemisphere medial identity-function engenders the conscious sense of loving and seeking consummation through sexual expression, expressive communication, and emotional contact. The sense of requisite contact with other people, and the sense of ideal subjective potential, of confidence, are contained in this identity locus. It integrates the "I."

(3) The Self Function This nondominant hemisphere medial identity-function integrates the sense of "me." The experience of *me* is an experience of affective potential, and affective signals derived from

sensory indications of feelings in the interior of the body. The identity integration of *being* produces an affective coherence centered inside a coherent sense of body unity.

(4) The Object Function This nondominant hemisphere lateral integrator of identity forms a mental representation of reality, which is experienced as a form of consciousness within perceptual representations of inanimate and animate objects. This identity function represents object-self and object representations to consciousness as aspects of reality. Feedback from the use of muscles and joints combined with vestibular data is perceived as represented as postural body images. These are integrated with vision and form objective object-self representations, of which the agent function may make use.

The Ego Functions

The ego-functions are hierarchically subordinate to the identity-functions. Arranged from an upper level of conscious functions to lower levels that exist at the margins of neuropsychological processing, the ego-functions channel signals from the identity functions out of the prefrontal areas into the distributed cortical and subcortical areas that are synaptically linked to the identity zones. They also channel signals up to the identity zones. Ego functions are often hemispherically paired, with reciprocal inhibition determining which function will be activated and which inhibited. A conscious verbal signal code determines how mental system-consciousness selects and channels the neuropsychological follow-through. We generally experience the ego functions as such automatic verbal cues as "try" (agent), "love" (subject), "remember" (self), and "watch out" (object-self).

(1) The Emotional System The paradigm maintains that emotional organization is distributed between the hemispheres. The affective component of emotion has its roots in the nondominant hemisphere. It is, I believe, a response to the received possibilities of fulfilling an action-plan. This possibility is assessed through the detection of particular salient features necessary for completing the plan, and it is for this reason that relief and frustration are the major affects. The expressive part of emotion originates in the dominant hemisphere. I believe it is a signal to carry out the plan.

In this view, frustration or irritation is a signal of difficulty or sustained novelty in the pursuit of the appropriate signifiers. Affects of sadness or depression relate to the sense that the salient signifiers are not to be found in reality so that the action-plan must be relinquished. Another form of affective response, anger, is a communicative signal of displeasure (as in attacks on the nature of the other person's being, or in sore complaints about feelings of abandonment) that displays variants of frustration.

The nondominant hemisphere's frustration/relief system must be in inhibitory control of dominant hemisphere modes of expression, for each step of an action plan must be released from inhibition by the nondominant hemisphere inhibition before feelings of success and satisfaction can be signalled to the dominant hemisphere. Our consideration of the emotional system must therefore include the fact that the expressive and gestural communications engendered by the dominant hemisphere, which are meant to convince the gratifier to reciprocate consummation wishes, are different from the affective indicators of displeasure that are another part of interpersonal communication. In evaluating emotional process we must decide whether we are considering a dominant hemisphere movement toward gratification, or a nondominant hemisphere reaction to the presence or absence of salients for gratification.

(2) The Anxiety System It makes sense to think of the anxiety system as a distinctly different ego-function from the emotional system. Anxiety triggers automatic changes in the unfolding operation of action plans. We have posited a dominant hemisphere signal system that evaluates errors or mistakes in action planning, and that inhibits continued unfolding of the action plan given irreconcilable errors or difficulty in method or procedure in the search for consummation. Claustrophobic (action-blocking-anxiety) discharges into the voluntary muscular system produce a state of tension in the muscular system, especially the expressive muscles of the face, head, and neck, that feed back as an inhibitor of action. Continued pressure for consummation in the face of such dominant hemisphere regulated inhibition produces a rage-response. Essentially, this dominant hemisphere form of instinctual action discharge brings the whole stereotyped repertoire of voluntary motor system action (fight) into play at once to remove obstacles to gratification. The quality of rage is different from the destructive nondominant hemisphere feelings that may precede a nondominant hemisphere flight reaction.

The nondominant hemisphere form of anxiety function operates to produce a reorientation and reassessment of perceptual cues when reality fails to provide expected cues of gratification. When reorientation does not produce a new fix on available salients, then agoraphobic (reality-denying-anxiety) discharges may develop. This is a state of nondominant hemisphere induced readiness to flee or withdraw from inhospitable reality. Thus, nondominant hemisphere regulated anxiety produces a hypervigilance to reality cues, and a stimulation of the neurophysiological systems of emergency readiness to flee.

It seems that the ego-functions of emotion and anxiety together signal each of the four identity-functions. Emotion regulates the medial identity-functions. Expressive feelings signal the *subject* to continue or discontinue pressure for consummation. Affect response signals the *self* about the degree of likely frustration or lack of it in the pursuit of action plans. The complete emotion system, therefore, indicates the likelihood of pleasure or pain expected in the pursuit of action-plans. The anxiety system signals the lateral identity functions. Error anxiety signals the agent to evaluate again whether pragmatic pursuit of the action plan is possible. Novelty anxiety signals the object-self as to the presence or absence of the familiar salient features in reality. The complete anxiety system determines whether system-consciousness can maintain its planned work schedule.

(3) The Thought System The ego-function of thought is a means by which action plans are formulated and prepared to be put into effect. Thought has (1) a dominant hemisphere portion that consists of the formulation of motives and their induction into existing action plan formulas and (2) a nondominant portion consisting of the formulation of a group or gestalt of signifiers that have produced a satisfactory end to action plans in the past. The nondominant hemisphere portion of this ego-function organizes social signification. Thus, we may use the term *association* to refer to the dominant hemisphere activity through which motives are made meaningful as they are formed into coherent plans. In this context *ideas* are existing organizations of thought including social metaphor and social idiom capable either of carrying significance and salience or of being judged irrelevant. Associations may be considered a subjective attribute of thought, and ideas an objective attribute of thought.

(4) The Speech System The ego-function of speech has a dominant and nondominant hemisphere organization. Thought must be transformed into speech and language in order to reach the system of voluntary consciousness whence it becomes available to conscious reflection. The meaning of speech is composed in the dominant hemisphere and generated according to subjective wishes. The structure of language exists as an organized syntax in the nondominant hemi-

sphere. A sentence is thus a unit of a system-consciousness synthesis in which the dominant hemisphere speech system generates the action and motive of the sentence and the nondominant hemisphere organizes the sentence pattern. The dominant hemisphere carries out and coordinates the stringing of phonemes (meaningful sound units), morphemes (units of meaning), articulemes (action of verbalization) in the generation of the meaningful subjective portion of the sentence, while the nondominant hemisphere produces melodic emphasis, prosody, and affective cadence. It also appears to place idioms and other forms of introjected language into the sentence structure.

(5) The Action System There are two action systems included in the ego function of action. The dominant hemisphere action system is a pragmatic, voluntary action system imbued with subjective meaning. Gesture is the prototype of this action system. The organization of motor processes according to subjective plans is built up around organizing instinctual ensembles which are behavior tendencies toward gratification. Just as the speech system has a subjective generating component and an objective language component so does the action system. The nondominant hemisphere maintains a simultaneous gestalt of actions, end points that represent completed action. The nondominant hemisphere receives feedback from the joints and muscle spindles telling of the progress of the ongoing action. The whole postural image of the body is developed and maintained in the nondominant hemisphere. This allows for the development of imitative action, which is different from the pragmatic sequencing of the dominant hemisphere. This is a simultaneous gestalt of body part relations. Thus, the dominant hemisphere proposes and initiates action, and the nondominant hemisphere brings that action under the control of spatial representations. The two action systems are integrated by the bilateral supplementary motor areas (SMA). SMA coordinate volition and imitation independent of lateralization.

(6) The Energy System The mental signal code for the allocation of energy is a basic ego function that determines which zone is selected for further neuropsychological enterprise. Each medial and lateral identity zone is paired with its reciprocal hemispheric partner in forming a mutually inhibitory basis for energizing neuropsychological functions. Thus cognitive work alternates the available dominant hemisphere energy for action (effort) with the available nondominant hemisphere energy for reality assessment (concentration). Emotional energy signals are produced by the medial zones paired as an *esteem-sys-*

tem. Thus subject-esteem is a gauge of confidence in action plans ("I can do it."), while self-esteem is a gauge of entitlement to fulfill action plans ("You deserve that"). The verbal cues relate the energy available for insistent consummation (passion) on the dominant side, and energy available for overcoming frustration on the nondominant side.

Concentration fixes the salients of reality as the features of an object in the formation of a percept, or the features of an intellectual task in producing a coherent and logical abstraction. This set of nondominant hemisphere neuropsychological functions requires the distinction of a whole background from salient features of the background. Clearly the energy for this kind of work can be exhausted. So too the energy of effort refers to the amount of voluntary work that can be undertaken. The inhibition of irrelevant action is clearly an energy conserving mechanism. *Attention* is an energy function that refers to the ability to hold conjointly information from the two hemispheres, using the product of effort and concentration in the synthesis of a single product of consciousness. Thus attention refers to the ability to hold a number of items in potential consciousness during the span of immediate memory.

(7) The Memory System Memory is a composite of several different ego functions. Verbal cues are used to re-enforce the use of different ones of these ego functions. Many of the interrogatory words, such as "who," "which," "why," "where," "what," "how," and so on, invoke dominant or nondominant forms of recollection. Recollection includes medial emotional forms of long-term-memory and lateral cognitive forms of stored information. Other verbal cues such as "remember" or "learn" make use of actually designated ego functions in the selection of neuropsychological processes to be invoked.

Immediate-memory (attention) refers to the items that can be held together in system-consciousness while they are being organized into a product of system-consciousness. *Short-term-memory* is the process by which the products of any 15 to 30 second duration of system-consciousness operation are fixed and demarcated for potential storage. Products of system-consciousness are *learned* as a variance from preexisting programs. Pre-existing programs are cognitively and emotionally re-enforced action programs with expected salients that are keyed to the identity requirements of a particular life period. When the unexpected action or reality happens new information is available as a potential alteration of an action program, and is so available to be learned. *Intermediate-term-memory* refers to the process taking 5 to 6 hours during which new information—products of the system-consciousness—are learned and retained in storage. Verbal memory learned as the memory of meaning, as well as pragmatic action memory is stored as dominant hemisphere information, while reality memory experienced as visual or perceptual data is stored as nondominant hemisphere information.

Generic-memory refers to the storage of emotional organizing memories. We must distinguish between conditioned emotional information which is stored within a combined cognitive emotional framework during a particular life period, such as "stoves burn," and the organizing emotional framework for the period. Each life period has its own generically organizing memories that we may call *basal-memories*. Such memories are placemarkers in our life experience, and their recollection orders personal time.

Generic memories are also organized hemispherically. Thus affective reverberation, giving rise to recollected, and often traumatic scenes, is a nondominant form of emotional memory. Traumatic events evoke strong affective memory formation, with fixed perceptual experience that remains a focus of long term memory formation. On the other hand, dominant hemisphere forms of long-term-memory are evoked as repetitively experienced action and consummation patterns (fixations).

(8) The Perceptual System The perceptual system is another bilaterally differentiated ego-function. However, within this system is a relatively undifferentiated bilateral inner core of simple pain, organ feelings, taste and smell. Other sensory systems too have diffuse representation. Each sensory modality has a quality of sensory intensity (brightness or loudness, for instance) that it conveys through the reticular and thalamic core and through other primitive systems, such as the "secondary somesthetic cortex" in which both sides of the body are represented as a global whole. Such sensory receptive areas are doubtless under more medial regulation by the distributed cortex of the prefrontal system, which means that each side of the brain is capable of forming some impression of the whole body. These phylogenetically ancient secondary sensory systems function outside of consciousness, except perhaps for the first few weeks of postnatal life, before the specific sensory systems come on stream.

The perceptual system that we consciously know assesses specific cortical and thalamic sensory data. The nondominant hemisphere forms a perceptual world of reality representation based largely on visual data. The discrete fixations of the eyes in feature detection provide a steady background for the evaluation of reality. The premotor cortex for visual processing is an immense field of feature detection accomplished through the formation of fixation groups. The prefrontal cortex, which probably evolved from the premotor cortex, uses these fixation points to assess the features of reality. The nondominant hemisphere also accounts for the composition of object and self images which it bases on visual and other discrete spatial data, such as the localization of sounds and the somesthetic localization of the body surface. Other main tasks of the nondominant hemisphere are (1) the formation of the body image as a concrete set of entities corresponding to postural models, and (2) the recognition of faces. The nondominant hemispere carries out these tasks through a collaboration between the posterior parietal areas of sensory reception and organization and the prefrontal areas in which this assessed information forms representations that can be used to organized intrapsychic perceptions of reality.

The higher level generalizations of the concrete representational data of the world forms the basis for metaphor. Gestalts of gestalts, as it were, form hierarchical organizations of logical structure. Perceiving the relationship of the part to the whole represents the kind of thinking that the nondominant prefrontal hemisphere organizes. Thus, the ability to generalize from metaphors and idioms requires an intact functional hierarchy of the nondominant hemisphere.

The dominant hemisphere has a different kind of perceptual organization. Here the consciousness of perception appears to be formed through the awareness of sequenced kinesthetic impulses. These minute motor sensations are linked to auditory sequence. The result of this conjunction is that the prefrontal apparatus forms the basis of speech out of the sequencing of articulation movements and their correspondence with auditory succession. The awareness of the generation of speech is the dominant hemisphere's form of perception. The dominant hemisphere is equipped with a large premotor area for the assessment and sequencing of the voluntary movements of the organs of articulation as well as an active center for the inhibition of speech and action.

A Structural Paradigm Nosology

Following the structural paradigm, all psychopathological syndromes may be naturally categorized (1) according to which hemisphere initiates the pathological process, (2) according to medial versus lateral zone involvement in the pathology, and (3) according to the level of system-consciousness impairment of function. The first two factors may be called *horizontal factors*, and the latter a *vertical factor*. Here, the question to be asked is whether social functions, identity functions, ego functions, neuropsycholgoical functions, or neurophysiological functions are impaired. The level of epigenetic fixation or trauma is another vertical factor that must be assessed.

I would like to place this aspect of diagnostic nosology within a larger contextual framework for diagnosis. We must consider paradigm factors within a context of life stage development. First, then, a psychopathological syndrome must be related to a particular stage of life. Each stage poses inevitable dynamic conflicts that the action programs of that stage of life resolve. Because each hierarchical formation of system-consciousness supersedes previous formations, and in the process reorganizes social functions, identity functions, ego functions, neuropsychological functions, and neurophysiological functions, pathological syndromes may not be viewed as achieving a linear development throughout epigenesis. Rather, the chronology of a syndrome must be viewed in terms of stage entry, stage maintenance, and stage ending. A school phobia in preadolescence, and an agoraphobia in the adult may be analogous in terms of the hemispheric and zonal contributions to the pathology; but one must not presume a linear syndromatic development of a single syndrome in the individual, for each period of life has its own inherent dynamic conflict that must be worked out in action plans.

A second portion of the diagnostic context has to do with social relations. An individual's identity synthesis exists in a matrix of extended political, institutional, social, and familial ties. More than this, one's identity-synthesis is particularly reflected, mirrored, and maintained through the complementarity of one primary relationship. Disturbance or change in that relationship cannot fail to have profound effects on the intactness of each integrating identity function. Thus, social disturbances give rise to threats to the general integrity of identity functions, and interpersonal disruptions give rise to specific disruptions in the foundations of identity synthesis. The social and interpersonal disrupting influences give rise to stress syndrome manifestations, and the particular identity function disruption within the context of the particular life stage gives the syndrome its particular character.

A third aspect of the diagnostic context has to do with the individual's genetic predisposition to a particular kind of psychopathology. I believe that this disposition may be best understood within the framework of an evolutionary tendency to medial or lateral processing, and dominant or nondominant processing. Women, for example, tend to do more medial processing, men to do more lateral processing. Thus

I surmise that genetic factors determine that an individual is predisposed either to the dominant hemisphere action mode, which includes accentuated coping, growth processes, and creativity, or to the nondominant hemisphere reality-participatory mode which includes perceptual processing and adaptation to and identification with prevailing social forces. Another genetic factor relates to the quantity of drive neurotransmitter substance generally available. Like other mammals, human beings may be divided into high drive dominant types, and low drive submissive types. These genetic factors must figure prominently in the disposition to personality types as well as in the disposition to particular psychopathological syndrome.

Given this context for diagnosis, let us go on to a definitive nosology based on the horizontal and vertical factors. I contend that horizontal desynthesis of system-consciousness produces *psychosis*. The nature of the psychosis depends, first, on which hemisphere causes the desynthesis. I contend that dominant hemisphere psychosis (schizophrenia, schizophreniform) is precipitated by the deregulation of the dopamine system, while nondominant hemisphere psychosis (affective psychosis) is precipitated by deregulation of the norepinephrine system. According to this theory, deregulation of either system exhausts the neutralizing capacity of the serotonin system, and this, in turn, desynthesizes system-consciousness itself. This means that psychosis is a result of a deregulation of the neutralizing capacity of the serotonin system in the zones of medial processing. In all psychosis, the acute stage is accompanied by a disruption of the neutralizing capacity of the serotonin system in the zones of medial processing. In all psychosis, the acute stage is accompanied by a disruption in the sleeping process, and this strongly indicates a deregulation of the neutralizing, generalized inhibiting serotonin system.

All acute psychotic decompensation, whether caused by organic factors or psychological-social factors, produces the same deregulation. Although the pathological process is inaugurated within a particular identity zone, all of the major zonal integrations of identity functions are compromised in the course of the acute psychosis. In some situations, such as an acute reactive psychosis, in which the ability to integrate reality is compromised, a particular zone will be more sharply affected than others.

The vertical organization of the medial functions is a guide to the unfolding sequence of symptomatology in an acute psychosis. The prodromal symptomatology will vary according to the nature of the psychosis-triggering mechanism, but the period of acute deregulation will tend to follow a sequence attributable to a hierarchical deregulation of function. Thus, social interpersonal functions are deregulated first. Next comes deregulation of identity functions, with concomitant desynthesis of subject/self collaboration. This, in turn, produces a deregulation of the ego-function of emotion, and, as the deregulation approaches the neuropsychological level, drive organization itself becomes disrupted. The psychotic syndrome proper begins at this point as lateral functionality is disturbed. The main difference between the major psychoses is simply that the schizophrenic types of psychosis unravel the dominant hemisphere functional hierarchy, while the affective psychoses unravel the nondominant hemisphere functional hierarchy.

The neuroses are also accounted for by the paradigm nosology that categorizes them as anxiety-syndromes. This nosology relates these syndromes to the function of anxiety in each hemisphere. These syndromes involve lateral functionings in each hemisphere, for anxiety safeguards the integrity of lateral functions at all hierarchical levels. Thus dominant hemisphere anxiety syndromes produce claustrophobic symptoms that interfere with the capacity to maintain flexible planning and new goal formation. Dominant hemisphere anxiety syndromes disrupt lateral ego functioning to produce feelings of action-paralysis, speech-inhibition, and compromise formed symptoms of speech and action. Analogously, anxiety syndromes of the nondominant hemisphere disturb the ability to participate in reality and to organize the features and representations of reality. The mental representation of social institutions is organized through a high level nondominant hemisphere identity function that fits the object-self into a mental representation of the social context. This function is disturbed in nondominant hemisphere anxiety syndromes. This disturbance gives rise to much of the phenomenology of *agoraphobia*, for in it the object-self is felt to be in danger of losing its social context and its social definition—i.e. its identity integrating function.

Character diagnosis also clarifies through application of the structural paradigm. Character pathology is a distortion of character based on the life-long distortion of a single identity zone. This means that character pathology must be seen in the context of a social distortion that begins in infancy and that leads to distortion of identity functions. As long as the distortion is consistent with the expectations of the surrounding milieu, the impairment does not lead to anxiety or even noticeable behavior pathology. Let us consider character distortion in regard to each of the four extended systems that are subordinate to system-consciousness.

Narcissistic character pathology is a distortion of the lateral dominant zone in which subjective goal formation (e.g., agency) hypertrophies. The sense of identity as agent subordinates all other identity functions to an unusual degree. The aim of consciousness is the inordinate reflection of all goals in other persons. Other persons become extensions of subjective goal formation.

The inverse distortion of character for this zone produces the *avoidant character*. This character expects his or her highly narcissistic goals to fail, with the result that the person does not strive appropriately for mirroring of subjective goals. The character does not expect reciprocating action by others to implement his or her own goals.

The dominant hemisphere medial subtype of character distortion is *hysterical character pathology*. In this disorder the individual is filled with insufficiently goal-oriented consummatory wishes. The repeated failure of sexual and interpersonal consummation leads to dominant hemisphere claustrophobic anxiety that fails to prevent fight and rage reactions. These secondary reactions become part of character structure and spring up in place of consummation aims.

The *schizotypal character* is devoted to the expression of basal organizing fantasies that derive from earlier life periods of development. However, there is little expectation that the expression of these fantasies will lead to satisfaction. Even more than the hysterical character, the *paranoid character* maintains a distorted rage response that is characterologically fixed as a response to consummation wishes. Naturally, the expression of each type of character distortion leads to a social or interpersonal response that fixes the character more devotedly than ever.

Now let us consider the nondominant hemisphere types of character distortion. The *masochistic character* distorts nondominant hemisphere medial identity functioning. The masochist is full of affective displays of suffering that emphasize the pre-eminence of the self while the object-self function that emphasizes the social person is completely diminished. The masochist must rely on affective displays of suffering at the same time that he or she feels that reality is a completely unreliable source of gratifiers.

The *borderline character* displays extreme medial affective distortion. The self function that regulates affect is distorted and does not prevent the outbreak of flight and fear reactions, and they, in turn, disturb interpersonal relations. Indeed, the borderline relies heavily on angry affective responses. The borderline is constantly frustrated and resorts constantly to reality distorting tantrum behavior.

The *dependent character* minimizes his or her own affective responses. Affect is present, but is inhibited by a ready guilt that minimizes the display of affect response.

The *obsessive character* emphasizes and distorts lateral nondominant hemisphere identity function at the expense of nondominant hemisphere medial functioning. He or she pays too much attention to all of reality's features, and to all social signifiers without being able to decide which are most salient. The obsessive is preoccupied with adaptation to the exclusion of affective response.

The *passive-aggressive character* emphasizes the obverse side of the obsessive preoccupation with salience. The passive aggressive fails to participate in precisely those functions that have the most social relevance and the highest adaptional relevance. He or she does not feel that the object-self has any place in the social structure.

The *sociopathic character* grossly distorts the lateral nondominant hemisphere function of salience detection. In the search for pure satisfiers he or she destroys reality without guilt. Neither the object-self nor other animate or inanimate objects are vested with participatory relevance.

External and internal pressures for social compliance are major factors in the tendency to develop character distortion according to genetic predilections for a particular kind of system-consciousness processing. In the case of external pressure, social structure may actually favor certain forms of character distortion. A highly structured bureaucratic society, for instance, favors character pathology that emphasizes nondominant hemisphere distortions. In the internal case, a relatively unstructured society will tend to favor more subjectively oriented, and subjectively distorted characters. As society changes its own structure, previously *society-syntonic* forms of character structure may become *society-dystonic* and therefore diagnosable as diagnostic nomenclature changes to reflect changes in the prevailing modes of social organization. The diagnosis of what is perverse, personally and socially, is also a case in point.

Psychosomatic Disorders

The paradigm view of pathological mechanisms and their relationship to genetic determinants as well as to social compliance may be understood through discussion of the precursor functions to the identity functions of system-consciousness that develop in the first few months of postnatal life. By the second month it may be observed that drive states follow one another in sequence. Consummation, orientation, and attention states follow one another in sequences that must be determined by the length of time during which the dopamine, norepinephrine, and serotonin systems can maintain them. Each drive state regulates a set of autonomic nervous system functions. The interpostion of attentional states upon the consummatory and orientation states evidences a regulation of the two catecholamine drive states through a single mechanism. Thus, even in early life attention is accompanied by a slowing of pulse rate. During each of the three states, sensory modalities and sensory intensities habituate, and parental responses to the various drive states become part of the habituated imagery that comes under externally regulated autonomic control. Gradually, each of the three drive states becomes organized and hierarchically controlled by a portion of the prefrontal cortex. Parental conditioning of the drive states is one determinate of individual drive-type predilection.

This means that the roots of psychosomatic disorder go even further back into the prehistory of the system-consciousness than do the roots of psychosis and character pathology. Because *psychosomatic fixation* involves each organ's having an autonomic balance of sympathetic and parasympathetic regulation that determines its functional adaptation, this regulating tone is capable of some social compliance. The parental response to distress that appears during the course of the prototypical drive states is a factor in the development of fixation to psychosomatic responses. Conceivably those illnesses that hinge on autonomic or hormonal imbalance such as asthma, addictions, hypoglycemia, thyroid disbalances, and allergic states to mention a few, have predispositions fixated due to social compliance of drive states in the first few months of life. These early vulnerabilities may become part of the distributed neurophysiological propensities for particular identity zones.

DSM III Criteria for Schizophrenia

The *DSM III*, which is the current standard manual of diagnosis, lists several ego and identity functions in its approach to evaluating schizophrenia.

> Invariably there are characteristic disturbances in several of the following areas: content and form of thought, perception, affect, sense of self, volition, relationship to the external world, and psychomotor behavior. (*DSM III*, 1980, p. 182)

Let us discuss these functions serially as they are clarified by the paradigm for neuroscience.

(1) Content of Thought

The *DSM III* stresses the formation of multiple delusions that often center on such phenomena as thought-broadcasting, thought control, thought-insertion, and delusions of reference. These primary delusions are, in my opinion, a result of the failure of particular component zones to integrate identity consciousness. The sense of failing subjective integrity undermines the ability to synthesize the whole of consciousness as a unity. The failure to synthesize leads to an inability to locate the source and generation of thought. As the synthesis of system-consciousness fails, nondominant hemisphere identity functions are also compromised, and this produces feelings of 'phoniness,' false significance, and the sense that all perceived events relate to the object-self.

(2) Form of Thought

The formal thought disorder of schizophrenia is a disruption in the coordination of two dominant hemisphere identity functions: (a) goal formation, and (b) the ability to plan for consummation. The lack of identity-regulated, coordinated sequence in the creation of action plans produces a lack of coordination in the generation of speech which issues as a defect in associations. This associational defect is often filled in with primitive sexual or rageful themes. The sequence here is that action plans come to a halt so that further planning becomes impossible. When thought is blocked, it loses its content so that much of the content of speech represents groping for an expression of lost subjective feelings of identity. Recourse is often had to overgeneralization of subjective themes in the attempt to maintain some connection with overall subjective identity. These themes have to do with ideals of loving God and with subjective pre-eminence— grandiosity.

(3) Perception

The *DSM III* assumes that the characteristic auditory hallucinations of schizophrenics are one species of hallucination among a series that issues from the various sensory modalities. This may be a mistaken view. It is an open question whether hallucinations arise entirely from within the brain, or whether they make use of ongoing sensory stimulation, remaking it into a form of inner speech. As we know, speech itself consists of a kinesthetically guided sequence of meaningful phonemes, so it is by no means a purely auditory phenomenon. The hallucinating schizophrenic often begins by hearing the radio or T.V. talking exclusively about him- or herself; later, the machine seems to echo the person's thoughts. Clearly this beginning phase of the formation of auditory hallucinations is a result of the breakdown of identity functions that distinguish between object-self and other objects in the world. In my own clinical experience, the quality of schizophrenic hallucinations is usually a response to the unspoken action plan question "What should I do?" The answer, an action plan framed by the extremis of the subjective identity quandry is hallucinated: "Kill yourself," "Jump off the roof," "Get out of bed." Given the failing sense of subjective identity, the schizophrenic wants to know what to do to re-establish subjective pre-eminence in his or her intrapsychic world. The hallucination, then, is often a response to this kind of natural, survival-oriented action plan question. It is interesting to notice in this context that if hallucinators hear spoken language disassociated from thought, they do sometimes identify the speech as their own thoughts. In fact, it sometimes happens that as the acute schizophrenic process resolves, the hallucinations lose their sensory character and become part of the thought system again, so that they are again felt to reside within the head. Most relevant is the fact that the dominant hemisphere proposes actions or questions, and the nondominant hemisphere provides answers in terms of what are the necessary salients. This reminds us that generated speech and the language system are separate synthesizable subsystems of the ego function of thought.

(4) Affect

Actually, the emotional blunting of schizophrenia has more to do with dominant hemisphere emotional expression than it has with the nondominant hemisphere frustration-irritability axis. When hopes for interpersonal consummation blunt, expectations of sexual gratification recede leaving deep feelings of shame in their wake. Finally, silly laughter may develop. These inappropriate expressions appear to be primitive forms of emotional expression that are no longer inhibited by the medial subjective identity function.

The failure to synthesize that develops in schizophrenia produces a generalized nondominant hemisphere reciprocal inhibition. When affect discharges do appear in schizophrenia they are sometimes experienced (like hallucinatory thoughts) as outside of ordinary identity. The schizophrenic may therefore cry without feeling that the crying develops as an inner response to life. Often, a formal depression develops, possibly as a reciprocal nondominant hemisphere response developing gradually in the weeks after the acute episode.

(5) Sense of Self

The *DSM III* touches on the loss of identity functions we have been describing.

> The sense of self that gives the normal person a feeling of individuality, uniqueness, and self-direction is frequently disturbed. This is sometimes referred to as a loss of ego-boundaries and is frequently manifested by extreme perplexity about one's own identity and the meaning of existence . . . (*DSM III*, 1980, p. 183)

Using the structural paradigm, we can refine these observations. *Self-direction* refers to the *subjective* identity-function of goal-direction, whereas *uniqueness* refers rather to the objective identity functions of an affectively integrated self and to the uniqueness of the object-self as an individual in the world.

(6) Volition

> Nearly always there is some disturbance in self-initiated, goal-directed activity, which may grossly impair work or other role functioning. This may take the form of inadequate interest or drive or inability to follow a course of action to its logical conclusion . . . (*DSM III,* 1980, p. 183)

This the the *sine qua non* of schizophrenia. The agent identity function's failure to maintain regulation of action-planning leads inevitably to a regressive process and further deterioration in other mental functions.

(7) Relationship to the External World

The schizophrenic's autistic motives are regressive substitutes for the unattainable adult satisfactions and love relationships. This autism is an overdetermined withdrawal, for the changing inner world takes up the schizophrenic person's attention. Acute schizophrenia leads to preoccupation with achieving a redress of subjective integrity.

(8) Psychomotor Behavior

The voluntary motor system produces a disturbance analogous to that found in speech and thought systems. When the voluntary subjective system becomes blocked, the whole gestural system of motor expression becomes waxy, inappropriate, charged with mannerisms, grimaces, and stereotyped behaviors. For the voluntary motor system expresses the tone of its consummation intentions in the repertory of expressive gestures. The blocking in this expressive system leads to autonomic nervous system changes and supersensitivity in the dopamine system.

Thus, considering the range of phenomenology in schizophrenia leads us to conclude that an inhibition, distortion, and regression of dominant hemisphere identity functions and ego functions is a prominent cause of the symptoms and signs of acute and subacute schizophrenia.

Other Dominant Hemisphere Syndromes

The *DSM III* formulation of diagnoses does not lend itself to an organization of the anxiety syndromes according to hemispheric distinctions. For instance, it includes claustrophobia within the criteria for agoraphobia. Using the structural paradigm, I see these two forms of anxiety phobia as operating within separate hemispheres. Psychoanalytic theory still wisely considers anxiety an important contributor to the formation of neurotic disorders. It seems to me that when character distortion fails to prevent the outbreak of anxiety, then the diagnosis of neurosis is justified. Therefore my preference in discussing the dominant hemisphere form of anxiety syndrome is to combine the *DSM III* description of the histrionic personality disorder with the *DSM III* description of claustrophobic anxiety, to explain adequately the makeup of *anxiety-hysteria.*

Like schizophrenia, anxiety-hysteria is a dominant hemisphere initiated syndrome of failure to consummate love strivings. In this illness, the expression of love wishes leads to intense feelings of shame and humiliation, with intense inhibiting anxiety signals. These signals inhibit the complete expression of love to the extent that the imminence of emotional or physical consummation produces claustrophobic, paralyzing frigidity of sexual and expressive emotion. Fight anxiety discharges lead to the wish to force the lover to take over responsibility for consummatory goals. This is one reason for the development of rape-fantasies. The hysteric wants the lover to function as a subjective extension of identity. She (or he) feels that her (or his) own subjective means are compromised. In the dream life of hysterics the experience of lapsing confidence and blocked love is seen as a sense of falling, or of missing the train of excitement, as sexual themes are relinquished.

Although it is triggered in one hemisphere, neurosis, like character disorder, develops compliance in the identity-functions of the other hemisphere. The sense of subjective incompetence is reciprocated in the nondominant hemisphere with the sense of having a damaged body. Women experience men as liable to discover and increase the body damage. They seek admiration both as a way of receiving objective verification of their body's integrity, and as a reassurance that the man can indeed be expected to function as an extension of their own love wishes. The sense of physical imperfection cannot be long quieted, however, for it originates in the recurring subjective sense of incapacity to culminate.

Let us consider the *DSM III* phenomenological criteria for diagnosing the histrionic personalty disorder.

A. behavior that is overly dramatic, reactive, and intensely expressed:

(1) self-dramatization, e.g., exaggerated expression of emotions
(2) incessant drawing of attention to oneself
(3) craving for activity and excitement
(4) overreaction to minor events
(5) irrational, angry outbursts or tantrums

B. Characteristic disturbances in interpersonal relationships:

(1) perceived by others as shallow and lacking genuineness, even if superficially warm and charming
(2) egocentric, self-indulgent, and inconsiderate of others
(3) vain and demanding

(4) dependent, helpless, constantly seeking reassurance
(5) prone to manipulative suicidal threats, gestures, or attempts.

Paradigm analysis of these symptoms and characteristics clears up some ambiguity in the description. Category B phenomenology boils down to an attempt to force the lover to take over the subjective function of loving. Category A phenomenology show the effect of the blockade of consummation strivings in the hysterical person's intensfied craving for excitement, and emotional expresson. When this heightened intensity is blocked, anxiety fight discharges take place—not designed to destroy the object, but to force the object to make love.

Narcissism

The structural paradigm leads us to conclude that disorders triggered and formed in one hemisphere have more in common with each other than they have with disorders whose etiology is in the opposite hemisphere. Thus, we may consider narcissism as a prototypical dominant hemisphere character disorder capable of resembling many of the other dominant hemisphere syndromes. Individual pathology, after all, is not confined to a single syndromatic description, and character disorder is not pure in its phenomenology. What is more, the developmental level of fixation has much to do with the nature of the character disturbance. This is particularly true of narcissism which may be taken as a distortion of both the medial and lateral subjective identity functions. These functions we may realize from a consideration of epigenesis are particularly vulnerable to distortion during periods of rapid increase in their scope, such as occurs in the six month symbiotic stage of development, in the eighteen month period of formation of verbal consciousness, in the early oedipal period, and in the early and middle stages of adolescence.

Depending on whether the narcissist is fixated by a poverty of understanding and mirroring, or by an overindulgence of response to the merest wishes, the narcissist's fixation is either to the ideal perfection of wanting nothing from the object or everything. The prototype experience to this fixation occurs in the symbiotic stage of development, from three to six months, when consciousness is first being synthesized as a mental system. Thus, the impoverished symbiotic infant craves the confirmation of subjective identity, and remains fixated to the wish for the perfect mirroring that the infant is entitled to in the symbiotic period. The overindulged symbiotic infant develops a prototypical fixation to grandiosity, such that the mere wish for subjective being moves the world. As he or she develops, the narcissist's consummation wishes are all felt either as terribly dangerous cravings in the case of the impoverished narcissist, or else as grandiose necessities requiring complete participation by the object. In either case the action body, with its expressive gestures, becomes intensely cathected with the wish for reciprocal response.

Our paradigm explains *DSM III*'s phenomenological criteria for narcissism. These criteria are:

A. Grandiose sense of self-importance or uniqueness, e.g., exaggeration of achievements and talents, focus on the special nature of one's problems.

B. Preoccupation with fantasies of unlimited success, power, brilliance, beauty, or ideal love.

C. Exhibitionism: the person requires constant attention and admiration.

D. Cool indifference or marked feelings of rage, inferiority, shame, humiliation or emptiness in response to criticism, indifference of others or defeat.

E. At least two of the following characteristics of disturbances in interpersonal relationships:
(1) entitlement: expectation of special favors without assuming reciprocal responsibilities, e.g., surprise and anger that people will not do what is wanted
(2) interpersonal exploitativeness: taking advantage of others to indulge one's own desires or for self-aggrandizement; disregard for the personal integrity and rights of others
(3) relationships that characteristically alternate between the extremes of overidealization and devaluation
(4) lack of empathy: inability to recognize how others feel, e.g., inability to appreciate the distress of someone who is seriously ill.

All of the above phenomena belong to the category of excessive subjectification of emotional, interpersonal, and intrapsychic life. Point A emphasizes the exaggeration of the subject identity function. The subject is pre-eminent. Point B shows the libidinization and exaggerated emphasis on subjective ideals, for notions of beauty and ideal love are the quintessential expressions of subject identity functions, since they are the sensed ends that terminate the action program of consummation. Point C shows how truly the objective self is emptied of any real affect. It is simply a vehicle for directing gratification to the subject. Point D defines the narcissistic rage response. This is a fight response turned to the purpose of subjective gratificiation. The narcissist is vastly concerned with subject-esteem so as to be totally vulnerable to feelings of waning confidence, experienced as emptiness. Thus, the phenomenology of point E centers on the narcissist's subjective need to experience the world as the extension of his or her desires. The subjective state of humans other than the self is of no interest.

Nondominant Hemisphere Syndromes

This hemisphere maintains the relationship between the self as a person (object-self) and reality. Here, the reality drive organizes signification which it regulates through the deployment of affect, which monitors the person's ongoing relationship with and participation in reality. The nondominant hemisphere also binds and modulates stimuli into known patterns. To use the paradigm to understand nondominant hemisphere psychopathology, we must expand our understanding of the reciprocal mechanisms of the medial and lateral identity functions in this hemisphere. The ability to adjust self and object representation to accord with conditions in reality ontogenetically is a major requirement for adaptation. In practical terms this means that the person must be capable of changing one's mode of social functioning in response to new reality conditions. There are many indications that this adjustment is carried out through a process of internalization of social features of reality.

In hypothetical terms the paradigm accounts for the identity level processes of *identification* and *internalization* through which features of the object-self and the object are exchanged. The process of fusion between object-self and object representations must be considered a form of mutative exploration in adaptive identity change.

DSM III Criteria for Mania

Here is *DSM III* on mania:

> The essential feature is a distinct period when the predominant mood is either elevated, expansive, or irritable and when there are associated symptoms of the manic syndrome. These symptoms include hyperactivity, pressure of speech, flight of ideas, inflated self-esteem, decreased need for sleep, distractibility, and excessive involvement in activities that have a high potential for painful consequences, which is not recognized. (*DSM III*, 1980, p. 206)

These diagnostic criteria form a picture of nondominant hemisphere identity and ego functions operating at high intensity. Clearly, mania intensifies the search for salience while it produces a sense of imminent gratification. The paradigm holds that a pattern of dysfunction in the object-self leads to a breakdown of the ego functions the object-self controls. This is mania.

(1) Mood Change

The concept of mood implies an identity container that maintains a pressure of affect over a period of time (Jacobson, 1954). Mood is this sustained pressure. Mania accentuates the function of the medial self as an energy container that gauges frustration. In manic inflation, the self encompasses the world. Thus, the manic anticipates that he or she alone is capable of bringing about the world's gratification, and this accounts for the expansive mood. This egocentrism becomes infectious. Irritability sets in when it becomes clear that immense gratification is continually looming without actual consummation.

(2) Hyperactivity

Manic hyperactivity includes the invocation of schemes and end stage action plans that evoke the sense of possibility of success. These schemes involve strangers without regard to the strangers' social relevance. Manics overlook limitations of the object-self as they lose the distinctions of social boundaries so that their plans lack salient features of social signifiers. This manic lack of distinction is the sign of disfunction in the lateral nondominant hemisphere identity function. When agoraphobic anxiety fails to stem the unfolding of novel features sought as the end to action plans, manics pull any partner into their plans.

(3) Pressure of Speech

The qualities of the nondominant hemisphere language system condition manic speech which is more musical, more lyrical, more prosodical, rhyming, and governed by accidental "clang associations" than ordinary conversation. Affective reverberations (rather than relevance to action) govern this communication. So does the internalization of language. Thus, manics often sing ditties and chant songs. They intone what they have incompletely internalized. More involved in the internalization and mimicry of language than in the generation of meaning, the manic tendencied person is often interested in language, speaks many tongues, and has a facility for learning new ones.

(4) Flight of Ideas

Ideas are sets of thought organized around already formed connections that have a reality independent of subjective generation. Manics focus on this nondominant hemisphere half of the thought process. Overextension of their object-self leads them to a failure to inhibit irrelevant connections. This means that they draw false and irrelevant features into their own thought process. Thus the flight-of-ideas characteristic of mania is based on perceptual similarities, intermediate steps in the formation of a complete thought. The manic incorporates the intermediate steps so rapidly as to overlook them in forming thought and giving utterance to it. To give an example: a manic patient babbled about seeing me in a Korean restaurant. The intermediate ideas he left out of account included his perception that I was moving slower than he was, in fact like a snail—and the last time he had eaten snails was in a Korean restaurant. This rapid connection of features on the basis of word identity, metaphorical equation, and perceptual similarity is typical for mania, and shows the distortion of a nondominant hemisphere ego function.

(5) Distractibility

The manic appears to have a forced response to every feature of the environment. Unable to screen out stimuli irrelevant to the perceptual process, he or she suffers from a clear distortion or failure in nondominant hemisphere perceptual process. For, just as speech meaning requires the inhibition of nonmeaningful sounds, so also visual feature detection requires the inhibition of irrelevant features. As novelty detection anxiety fails to produce inhibition, the past kindles into the present so that memorial connections intermingle with present stimuli, much as they do in the inspirational phase of poetic composition. This inability to screen out irrelevant stimuli is the result of a failure to maintain ego boundaries between the object-self and the object. Any person becomes an *externalized object-self* in acute mania. All of the factors that lead to distractibility and confusion can be accounted for by a failure of nondominant hemisphere binding energy. Ordinarily, serotonin and GABA must inhibit the irrelevant connections, creating a context, while norepinephrine facilitates the relevant ones.

(6) Inflated Self-Esteem

The inflated self-esteem of mania shows a lack of integration in the identity function of self. The inaccurate signal of increased energy is conveyed to the object-self which feels able to encompass any external object and to assimilate alien objects to the self. False internalization can reach delusional proportions so that the manic may feel that he or she possesses the attributes, or even the flesh, of great figures in the world.

(7) Decreased Need for Sleep

Because in mania the nondominant hemisphere participatory process never shuts down, the manic barely sleeps. He or she maintains a false but constant contact with reality. Satisfaction is so imminent that sleep must be postponed.

Mania exaggerates the nondominant hemisphere ego and identity functions. Unconstrained by novelty anxiety, the manic's participatory process takes on inappropriate aspects of the real world in the attempt to force salient features into availability.

Major Depression

DSM III describes major depression as in many ways the inverse of mania. Actually, the depth of the hierarchical disintegration of functions goes further in major depression than it does in mania. Even psychotic mania provides a defense against disruption of most neurophysiological functions. Endogenous depression effects them deeply. According to *DSM III* major depression shows

> either a dysphoric mood, usually depression, or loss of interest or pleasure in all or almost all unusual activities and pastimes . . . symptoms include appetite disturbance, change in weight, sleep disturbance, psychomotor agitation or retardation, decreased energy, feelings of worthlessness or guilt, difficulty concentrating or thinking, and thoughts of death or suicide or suicide attempts. (*DSM III*, 1980, p. 210)

Let us apply the paradigm to these diagnostic criteria.

(1) Mood Disturbance

The depressed person's dysphoric mood, loss of interest in life, and despair indicate less that action plans have no consequence and more that he or she feels they are irrelevant. This mood of irrelevancy interdicts the nondominant hemisphere's overriding identity integration—its participation in reality. The identity function of the indifferent self is gravely reduced.

(2) Appetite Disturbance, Change in Weight, and Sleep Disturbance

Together with such other autonomic disturbances as dry mouth, constipation, and night sweats, disturbances in appetite, weight, and sleep represent the organism's adaptive response to the loss of hope in the fulfillment of future action plans. Reduced involvement in reality accompanies indifferent hope. Reduced involvement in reality shows up as neurotransmitter resources are exhausted. The hormonal deregulation must also be seen as a conservative trend preceding a waiting period in which the depressive pauses to see if conditions in reality will change sufficiently to allow the redevelopment of future-oriented action plans. The observed tendency to limit the periods of deepest mourning to about 2 months and the fact that some periods of depression appear to be self-limiting indicates that a new balance of enhanced neurotransmission eventually restores hope to the system so that interest in reality and in other persons may again appear.

(3) Loss of Interest or Pleasure

Of course the loss of interest in life must be a result of the diminution of available neurotransmitter and neuromodulator induced energy to engage life. The zest for life and for new experiences, such as we see in travelling, seeing new places, and experiencing works of art, indicates that the exploration of reality forms new salience that enhances neurotransmitter resources in the long run.

Thus we must include in our understanding of the nondominant hemisphere process that the formation of new salience allowing the binding of the working energy of system-consciousness must yield a pleasure signal. New experiences are all treated as without interest by the depressed person.

(4) Psychomotor Agitation

The voluntary motor system does not produce the psychomotor agitation of depression. This agitation represents disaffection and irritability with objects as they are. It is a physical sign of the depressive's tendency to obliterate and deny offending reality. We must be aware that our sense of the reality of objects is formed out of somesthetic images fed back from the joints and muscles, in conjunction with visual images of objects. The motor irritability may well be traced to a sense of irritation the depressive has with the physically composed images of the object world.

The motor agitation of some depressives announces that no action can lead to gratification. Other depressives' retardation of action and speech is also based on the feeling that no action will result in accomplishment. No ideas are new. The depressive feels that every action and every speech has been tried and found to be unavailing or untrue.

(5) Decreased Energy

The feeling of decreased energy may be taken as a direct consequence of the lower available energy levels. All of the drive neurotransmitters and all of the relevant neuromodulators appear to be diminished.

(6) The Sense of Worthlessness and Guilt

The upper level identity functions of the object-self are related to socialization. In depressives these are impoverished through a lack of invigorating energy. The depletion of neurotransmitter announces itself in the lowering of self-esteem and in a signal of guilt, for guilt is an affective mechanism for shutting down activity related to social reality. It is the means by which the intrapsychic system signals that action plans must be vastly curtailed; and it is to the sense of object-self identity as feelings of worthlessness are to the self identity. In the acute stage of depression, guilt is a constant reminder that social interaction is not allowed. Continuing to search for salience in the face of impoverished neurotransmission and supersensitive noradrenergic and serotonergic receptors can only lead to further pathology.

(7) Difficulty in Concentration

Concentration must be a nondominant hemisphere ego mechanism, for this mechanism involves binding salient reality and excluding nonsalient reality features. When there is too little energy available to form representations of the features of reality indecision follows. With decreased drive energy the depressive is unable to evaluate salience in reality.

(8) Thoughts of Death or Suicide

There are at least two causes for the death wishes of depression. In the first case, the depressive seeks death because it extinguishes the affective pain that accompanies the perennial giving up on action plans. This kind of death wish is about doing away with a "useless self." In the second case, the depressive seeks to attack an internalized object that fails to mediate action plans and to make satisfaction possible. An old pathological identification is at issue here. In pathological identification with a lost or newly nongratifying object, the object is taken into the object-self as an antipathetic, incompletely internalized structure. The suicide that may follow in the wake of pathological identification represents the destructive force of guilt directed against an object-self that is felt to have become socially unworthy because of the indwelling pathological object.

Psychotic Features in Affective Disorders

Psychotic features are delusions and hallucinations that sometimes accompany the other symptoms of major depression or mania. The psychotic features that are seen in mania and depression may contain the same primary delusions that we see in schizophrenic illness; however, in mania and depression these features bear the imprint of the nondominant hemisphere. When the psychotic features are *mood congruent*, i.e., tending to re-enforce the primary affective symptoms, they contain nondominant hemisphere identity themes. Guilt, suffering, and self-degradation are felt to be imposed from outside ordinary identity. Still, these mood congruent psychotic features re-enforce identity functions by emphasizing identity preserving mechanisms of guilt and denial that inhibit useless endeavor and pain. When, however, the delusions become so generalized as to be no longer mood congruent, then a fragmentation of identity functions and ego functions begins.

Other Nondominant Hemisphere Syndromes

Agoraphobia

Agoraphobia is a condition in which the self is alone in a world of strangers. Novelty is signalled by being alone. The anxiety is triggered by the fear of the object-self losing its social standing. It is an anxiety about loss of mediated relationships to the social world. Anxious withdrawal from the social world leads to clinging behavior with one or two important objects. This behavior induces a rejection. The sense of being abandoned leads to a withdrawal from one's own object-self identity function. Agoraphobia is a self-fulfilling prophecy as the anxious clinging behavior does lead to social isolation. It is readily apparent that the agoraphobic syndrome can be a prelude to affective syndromes of all kinds.

Masochism

If hysteria is the syndrome of subjective (dominant hemisphere) medial failure to love, then masochism is the syndrome of objective (nondominant hemisphere) medial failure to ask for it. In masochism, the medial affective self is felt to be ineffective. Affect discharge is not felt to communicate. In the masochist's perception, only a state of suffering beyond affect has communication value. Feeling that they have no self presence in the world, the masochists leave their own self out of account in social dealings. Self-esteem is so chronically low that masochists deny its depletion. They use their reality binding energy to emphasize object functions and to de-emphasize object-self functions to the point that the object-self exists as a suffering, sacrificed, disadvantaged entity.

The lack of object-self standing comes about as a distortion of character due to parental failure to objectify and mediate the relation of the masochistic child to the world. The parents of masochists are often overwhelmed and burdened by the world. Under such conditions, a 'good' child does not display affective need, and certainly not any destructive anger, for that only adds to the parental burden. The good child is altruistic and selfless. Her or his organizing fantasy is some stage-appropriate variant on the theme of the parents' ultimate recognition of the sacrifice of the self. In this fantasy, parental love will one day re-enter the vacuum of the affective self and compensate for lifelong frustration. The characteristic anxiety of the masochist is that even complete self-sacrifice will result in no response. The masochist teaches us that feelings of involvement with others are based on empathic responses to others' pain, as well as involvement with others' pleasure, for the masochist cannot resist the pain of the externalized object-self projected into another person.

Obsessive-Compulsive Syndrome

Agoraphobic anxiety is at the center of obsessive compulsive character distortion, just as claustrophobic anxiety is at the center of the dominant hemisphere hysterical syndrome. The obsessive fears that his or her aggressive behavior in attempting to alter reality and the social world will result either in destruction of the object or in destruction of the object's love. Both outcomes will leave the person deserted. In order to defend against his conflictual state centered by agoraphobic anxiety, the obsessive indulges in intense feature-detection. Obsessives are concerned with their own, and others' significant features. They believe superstitiously that the identification of such features will mean success or failure to their action plans.

As a nondominant syndrome, the obsessive-compulsive syndrome emphasizes the objective aspects of identity in a way that is out of proportion to the subjective aspects. The obsessive is more interested in social language than in his or her own generated speech. Forever in doubt about meaning, obsessives sense that they are not generating meaning on their own. Because they are convinced that satisfaction must be wrested from the world by reading the world properly, they try to make other people extensions of their own object-self. They try to make others responsible for decision making and for feature detection. Subjective pleasure and initiation become secondary to aggressive, reality-oriented conflicts.

Borderline Syndrome

The borderline syndrome is a nondominant hemisphere lack of maturation of the capacity to deploy the reality drive which cannot be used in appropriate affective response or in appropriate exploration of reality. Borderlines show an exaggeration of the masochist's medial selflessness. They have no appreciation of the efficacy of the self in regulating affect. They feel no self-esteem. Instead, they display all or nothing aggressive affect—reality destroying anger or indifference. The lack of medial development produces a compensatory concentration on object-self and object development. The lack of effective medial self makes others largely equivalent. The borderline object-self is treated as no more or less significant than objects. Both are unstable and volatile, largely arbitrary centers of aggressive power. Thus, the borderline object-self is a *false object-self,* easily changeable, and no more valuable than the interchangeable objects of the borderline's world.

The structural paradigm is useful in aligning the phenomenology of the borderline syndrome. According to *DSM III*

A. At least five of the following are required:

1. impulsivity or unpredictability in at least two areas that are potentially self-damaging, e.g., spending, sex, gambling, substance use, shoplifting, overeating, physically self damaging acts
2. a pattern of unstable and intense interpersonal relationships, e.g., marked shifts of attitude, idealization, devaluation, manipulation
3. inappropriate, intense anger, or lack of control of anger, e.g., frequent displays of temper, constant anger
4. identity disturbance manifested by uncertainty about several issues relating to identity, such as self-image, gender identity, long term goals or career choice, friendship patterns, values, and loyalties
5. affective instability: marked shifts from normal mood to depression, irritability, or anxiety, usually lasting a few hours and only rarely more than a few days, with a return to normal mood

6. intolerance of being alone, e.g., frantic efforts to avoid being alone, depressed when alone
7. physically self damaging acts, e.g., suicidal gestures, self-mutilation, recurring accidents or physical fights
8. chronic feelings of emptiness or boredom.

These collective phenomena are almost purely nondominant hemisphere deregulations of identity functions, ego functions, and neuropsychological functions. This is a syndrome of holistic instability of the nondominant hemisphere, just as narcissism is a syndrome of instability of the dominant hemisphere identity, ego, and neuropsychological functions.

Taking the phenomena up in order we see that the unpredictability in (1) relates to an attempt to wrest a missing satisfaction from the object world. This is not so much a true impulsivity as it is an attempt to force an end to frustration. The unstable interpersonal relations and shifting anger, irritability, and depression (2, 3, and 5) all show affective decontrol and failure of the self to function at a high level of identity regulation. The identity (4) appears to lack integration of the various aspects of representation. We have seen that according to structural paradigm personality representation is an ego function of the nondominant hemisphere's lateral zone. The attack on the "self" (7) shows instability of the object-self identity components that organize *object-self images* and *object-self representations.* Finally, the chronic feelings of boredom and emptiness (8) show the self and object-self as lacking in integration all up and down the vertical hierarchy of functions.

Sociopathy

In paradigm terms, sociopathy is a character disorder in which the lateral object-centered identity functions fail to develop normally. Sociopaths have little sense of medial affect, and even less sense of its real impact on others. In their perception their own object-self is real, other persons are not. They therefore mimic and ape others; neither truly identifying with them, nor internalizing their character structure. Because other people are not real to them they fail to distinguish the real significance of social communication. They feel that they have the right to manipulate reality and other persons in order to produce the combination of features that will supply them with what they wish. The logic of the sociopath is a set of integrated, but not internally consistent formulations.

MIND/BRAIN LOCALIZATION

Let us briefly review our map of mind/brain structural-functional congruences in preparation for a reconsideration of some relevant neuropsychiatric syndromes. The frontal areas maintain, regulate, and control cortical functions via layers of selective inhibition that are hierarchical. The dominant hemisphere selectively inhibits action schema, hierarchically organized as plans, goals, pragmatic sequences, and action (kinesthetic) images. Damage to the dominant prefrontal area produces damage to impulse control. At the dominant premotor level, sequencing involves the ability to string expressive communication in the right order. Articulation and the smooth flow of expressive communication depend on the intactness of this area. The posterior dominant hemisphere receives kinesthetic, auditory, and visual information and produces the necessary inhibition of irrelevant information to form phonemes and visual morphemes necessary to receptive hearing and reading. The posterior dominant medial temporal area of cortex produces the sense of *meaning* that is related to dominant hemisphere forms of memory. The dominant hemisphere stores memory of action and speech schemata that have produced meaningful resolutions of consummatory needs in the past. The prefrontal assessment of the sequenced verbal memory necessary to meaning constitution provides the dominant hemisphere with a sense of the future. The fulfillment of action plans requires the sense of meaning to be satisfied. Thus, the dominant hemisphere must ascribe meaning to salient features before the limbic signal is given to proceed to the next step of an action plan.

The nondominant hemisphere selectively inhibits irrelevant background features in forming relevant feature groups. This hemisphere renders images and representations of space and of the body as object entities in space. The nondominant hemisphere lateral frontal areas maintain the ability to imitate through equating muscle positions with objects. The nondominant medial area of the prefrontal zones regulates affect discharge. Damage to this area is one cause of affective lability (lack of modulation). The nondominant posterior parietal areas maintain the ability to produce groups of object features that are kept categorically separate among simultaneously presented informations. This part of the nondominant hemisphere distinguishes phrases such as 'my uncle's brother,' from 'my brother's uncle.' The posterior part of the nondominant hemisphere maintains the body image as a whole entity in space. The temporal area of the nondominant hemisphere produces the feeling of familiarity, which is a major memory function that involves the ability to take a set of features from the present and relate it to the past. Object and spatial familiarity is produced through this comparison (salience). Long term memory (including memory of visual scenes with affective reverberation) is organized in the medial temporal area of the nondominant hemisphere in connection with the limbic amygdala of this side. Thus, this area adds the quality of familiarity to the perception of salience. In order to proceed with action plans, the system-consciousness must judge that the salient feature is both meaningful (good) and familiar (true). Thus, the entire cortex comes into play in the ongoing work of system-consciousness.

NEUROPSYCHIATRIC SYNDROMES

Together with various generalizations derived from cases of cortical damage, the paradigm allows us to formulate principles helpful in approaching the phenomenology of the neuropsychiatric syndromes. By neuropsychiatric syndromes I mean those mental syndromes whose etiology is purely organic. The effect of massive strokes exemplifies one principle. The hemisphere opposite the afflicted one shows a reciprocal response. Thus, massive nondominant hemisphere damage can lead to a dominant hemisphere engendered reciprocal response of *apathy*. Presumably when the dominant hemisphere is overstimulated and released from inhibition, its resources are exhausted producing a state where no incentive for action planning or goal formation can be engendered. Similarly when massive dominant hemisphere strokes occur, the opposite nondominant hemisphere appears to exhaust its affective resources, ending in a state of *indifference*.

Attention to the functional hierarchy allows us to 'localize' mind/brain dysfunction. Higher level ego functions can be distinctly delineated as disturbed within a particular zone of identity integration. Thus goal orientation and flexibility in changing goals may be considered ego functions subsidiary to the identity function of agency. Maintenance of a coherent sense of reality, the ability to imitate and reproduce reality accurately, and orientation for time, place, and person are all ego functions related to the identity function of the object zone. Such higher level ego functions as cognition are dependent on the intactness of both the agent and object zones of identity integration.

A consideration of the lower level of system-consciousness synthesis takes us into the realm of neuropsychological functions. The ability to summon consciousness, to maintain attention, to make efficacious effort and to concentrate, are all lower level neuropsychological functions that are synthesized at the neurological margins of the system-consciousness. The immediate memory function of sustaining a sufficient number of items of information in common linkage is a basic low level system-consciousness ego function. These items in turn are formed out of the neuropsychological ability to link morphemes and words sequentially, as well as the ability to join object features in simultaneous linkage to form percepts. Interruption of these neurological margin functions gives rise to the fluctuations in attention termed delirium.

Temporal Lobe Epilepsy

The paradigm helps explain the effects of temporal lobe epilepsy and ictal phenomena in each hemisphere. Posterior medial temporal drive image zones are linked to prefrontal medial cortical discharge areas. In normal functioning, images of salient goodness and trueness processed in the temporal lobes give the signal of discharge readiness that triggers the completion of an action program. The prefrontal system-consciousness assesses these signals and activates the discharge process. In temporal lobe epilepsy, however, rather than produce a signal to prefrontal system-consciousness, the temporal (amygdalar) signal of discharge readiness sets off a spreading seizure phenomenon itself. The brain's intrinsic mechanism for producing program completion is compromised at a vital moment.

The compromise has an interesting psychological effect. The fact that the signal for discharge sets off seizure activity and a lapse in consciousness rather than the psychological experience of satisfaction and relaxation makes the seizure itself a symbolic equivalent to satisfaction. Inasmuch as the seizure phenomenon may consist of primitive behavioral sequences that the onlooker may equate with gratificatory behavior, the phenomenology of the seizure also receives social re-enforcement as a symbolic gratification. The combination of false gratification and social responsiveness to the false gratification, as if it were in some way real, produces a *hysteroid* personality structure. For the psychological character structure of hysteria is such that the hysteric also gets close to gratification but then backs away, because of anxiety, finally producing a symptomatic compromise form of gratification. The hysteroid cast to the personality in temporal lobe epilepsy shows that lower level neurological disturbances result in identity function adaptations. I believe, indeed, that every neuropsychiatric syndrome, (indeed many of these syndromes may be very mild and subclinical yet capable of exerting an organizing influence) produces an effect on the overall integration of identity functions. Many of the resulting identity function distortions become socially significant in the person's approach to life. Temporal lobe epileptics, for instance, are stigmatized socially and traumatized personally in such a way that the illness becomes a significant factor in their identity construction. This is clear in the writing of Dostoyevsky, which is full of the imagery and actual indications of temporal lobe epileptic aura of sought gratification transformed into a loss of consciousness.

The hemispheric distinctions between dominant and nondominant temporal epilepsy show up not only in the higher incidence of schizophreniform psychosis in the period between seizures in dominant hemisphere epilepsy, and in the higher incidence of affective psychosis in the nondominant hemisphere temporal lobe epilepsy in the period between seizures (Trimble, 1980), but also in the characterological type of the epileptic. Thus the imminence of greatness felt by the dominant hemisphere epileptic, sometimes shown in the hypergraphia, relates to the dominant hemisphere assessment of expectations of satisfaction, which are forever imminent but never released. Dostoyevsky's writing reflects this sense of imminent achievement. The continuing failure at program completion leads to the deep gloom and morbidity (perhaps a nondominant hemisphere response to the disrupted dominant hemisphere function) that is so clear in a book like *Notes from Underground.* The title itself expresses the theme of a signal code that can not reach fruition.

Dostoyevsky also dramatizes the depersonalization and derealization that are common to temporal lobe epilepsy. The inability finally to consummate leads the individual to conclude that there is something wrong not only with the experience but with the experiencer. The search for consummatory greatness, the feelings of derealization, and the image of literal split brain identity functions are evident in Raskolnikov's crime in cleaving the old woman's skull in *Crime and Punishment.*

Recently, our notion of the phenomenology of temporal lobe epilepsy has expanded. Remillard et al. (1982) have investigated sensual phenomena leading to orgasm that can be an accompaniment of temporal lobe epilepsy in women. They found a preponderance of right amygdala discharge accompanying orgastic and sensual phenomena in temporal lobe epilepsy. These discharges were more likely to occur in the premenstrual period. It appears likely, then, that estrogen and progesterone act as neuromodulators in the amygdala area. Heath (1972) also found the right septal and amygdalar nuclei discharging during orgasm in epileptic patients. This suggests that the right hemisphere is responsible for the release of orgasm. This information dovetails with the fact that manic induction begins with a right hemisphere triggered blank dream and orgasm. Structural paradigm takes the right hemisphere as inhibiting discharge processes until a clear signal of readiness is provided by the dominant hemisphere.

The use of deep probing electrodes in cases of temporal lobe epilepsy can give a very direct testimony to the differential nature of brain areas in function. One case in particular, documented by Gloor et al. (1982), deserves consideration. It is instructive because it documents the manner in which traumatic experience spreads to kindle a wide gamut of long term memory experience and therefore to organize expectations of salience and to organize defenses against a repetition of trauma. In this case, an eight-year-old boy was forcibly held under water by an older boy. The younger child apparently sustained damage by anoxia to precisely the portion of his brain that was most stimulated by the intense fear for his life—the right amygdala. Shortly after the incident, the boy developed temporal lobe seizures that were ushered in with complex visual hallucinations, auditory-visual hallucination, fear and déjà vu. When, as a twenty-two-year-old man, this epileptic's right amygdala was stimulated, he described his experience of the past distinguishing it from the present in the following manner:

> "A kid was coming up to push me into the water. It was a certain time, a special day during the summer holidays and the boy was going to push me into the water. I was pushed down by someone stronger than me. I have experienced the same feeling when I had petit mals before.". . . When questioned, he said that this had been a true event in his life which occurred when he was about eight years old, probably shortly before his seizures began. (Gloor, 1982, pp. 134–135)

The patient experienced the focal past event with compelling interest at the same time that he was able to disengage from it and to report it as an experience still detachable from the present. When, however, the right amygdala was stimulated at higher electrical intensity, the experience was more of a forced memory that was less distinguishable from the present.

> "It was out by the sea and high up on a cliff, a feeling as if I were going to fall. It was a scary feeling. We are there, a world within a world, all of us were there. It is so real, yet so artificial." (p. 135)

When questioned, the patient related the scene to a picnic he had attended with another child (the girl next door) and her parents. The scene is markedly similar to the description of the drowning experience. The original trauma organized like perceptual-emotional events, linking them together memorially. Let me explain: I attach importance to the patient's use of "we". *We* includes the experimenters. As the young man's memorially relived fear increases, his ability to separate the present from the past diminishes. The feeling of falling and being pushed down is an image of entrance into the epileptic aura, an image of being forced into the experience by the examiner. It revives the original trauma of being held down under the water. This coalescence of experience around the organizing imagery of what is traumatic or fixating reveals the strength of the kindling experience to form new organizing mental structure.

> Later during the day the right amygdala was again stimulated with a higher intensity of 3 mA. He again became nauseated, felt that he was somewhere in the country with Tracy at a place where he had been before, and felt that it was dark and raining. He was extremely frightened and pale, and pleaded not to repeat the stimulation. After-discharge was present in the limbic structures and the neocortex of the right temporal lobe. (p. 135)

This patient's experience shows us that as the traumatic experience organizes later experience, it also defends against it by producing anxiety that warns of imminent danger. In the memory the young man is again in a scene of fear, now with the protective female, but he is still exposed to the water and to the feeling of falling. Here the nondominant hemisphere organizes the patient's participation in reality and his feeling of familiarity. His feeling of going beyond anxiety either to a seizure or towards the original trauma produces a beginning disintegration of his objective identity functions.

The examiners gave one further electrical stimulation, this time to the right hippocampal area.

> "I am balancing on the edge of a fountain. I have often experienced this in 'petit mals'. It is like I am in an old story book. I am afraid to fall into the fountain." He smiled. When asked why he did so while claiming to be afraid, he said, "Because I have experienced this so often." After-discharge occurred in the right parahippocampal gyrus and the right hippocampus with some spread to the right amygdala and the right temporal neocortex. (p. 136)

Thus, falling into the past, losing the sense of identity, and developing a seizure are all aspects of this patient's variety of temporal lobe epilepsy. This form of temporal lobe epilepsy must remind us, if we have seen psychotic 'angel-dust' users, of their forced memory and entrance into their own traumatic past. They take the drug to remove themselves from their present reality and end up in a complete contact with the original trauma they sought to avoid.

Chapter 3

THE PROCESS FOR NORMAL CHANGE AND THE SCHIZOPHRENIC PROCESS

Chapter 1 outlined seven epigenetic stages in the process of intrapsychic change (pp. 45–46) Looking at this process in a more experiential way, this chapter condenses those seven stages into five by putting together the first two stages, and then the third and fourth stages as they appear in Chapter 1. Essential here is that in adult life there are in fact two main kinds of normal psychological change. I call one the process for psychological growth. It is a creative process undertaken by the dominant hemisphere. I call the other process psychological adaptation. The nondominant hemisphere initiates it. Adults undertake both kinds of normal change when reality presents them with irresolvable difficulties. The present chapter looks at the process for psychological growth.

THE PSYCHOLOGICAL GROWTH PROCESS

Stage I: Identity Challenge

Social relevance and appropriateness govern the formation of action programs that regulate daily activities. The tested action programs regulate ordinary problem solving and decision making that each person develops in his or her own social context. System-consciousness change begins when social disruption and failure in basic identity-keyed programs trigger a change in drive pressures. By identity-keyed programs, I mean those action patterns that organize work, love, and family life.

When these programs fail the individual experiences repetitively signalled error and novelty anxiety. These signals evoke long term memories, specifically those memories that provided the experiential basis for the formation of *identity-constructs* that have regulated mental life in this particular era. Preoccupation with identity is thus a critical aspect of the traumatized mental response that leads into the regressive aspect of the process for system-consciousness change.

The essence of this traumatic reflection is a direct but neutral experience of psychological and social identity. The neutralization of all identity components is the prelude to the disruption of their synthesis. When identity functions themselves are compromised, the stability of the synthesis of intrapsychic process in imperiled. In identity-trauma all four identity integrations are compromised as a person feels (1) unable to generate impact (loss of subject), (2) unable to communicate (loss of agency), (3) unreflected (loss of self), and (4) not substantiated in reality (loss of object-self).

Diverting drive energy to higher level identity functions depletes the ego and neuropsychological functions. As neuropsychological regulation begins to fail, so do everyday means of problem-solving. Here is Janis and Mann's (1977) description of the state of identity challenge:

> *In a severe decisional conflict, when threat cues are salient and the decision maker anticipates having insufficient time to find an adequate means of escaping serious losses, his level of stress remains extremely high and the likelihood increases that his dominant pattern of response will be hypervigilance* . . . A person in this state experiences so much cognitive constriction and perseveration that his thought processes are disrupted. The person's immediate memory span is reduced and his thinking becomes more simplistic in that he cannot deal conceptually with as many categories as when he is in a less aroused state. (p. 51)

Clearly, the trauma of impossible decision making in extreme threat completely facilitates the norepinephrine and dopamine systems as well as the serotonin system.

The overenergized system-consciousness can no longer synthesize information in an effective way. Immediate memory operation (i.e., attention to signed valence) is disrupted. The onset of perseveration indicates the failure of subjective action schema (propositions of action) to be matched with reality features. This breakdown in functional synthesis triggers further disruption in the brain/mind apparatus ability to regulate and select.

Stage II: Regression—Universal Trauma

At this stage, both fight and flight prove fruitless, and this initiates a prolonged neurophysiological response of massive anxiety or panic. At this point hierarchical deregulation of the identity, ego, and neuropsychological functions coupled with neurophysiological activation occurs:

> The hippocampus, with inputs from other areas of the frontal cortex, appears to regulate the activities of the corticotropin-producing hypothalamic cells when control of territory has been lost or when old patterns of responding fail to produce rewards. (Henry & Ely, 1980, p. 91)

Like the animal that fails to maintain its territorial status, the human who can no longer make use of his or her environment reverts to a limbic hippocampal response that releases CRF–ACTH production and somatostatin-growth hormone from inhibition, which produces a prolonged stress response.

Several experiments lead to this conclusion. When Mason et al. measured human stress response to expected and unexpected trauma, they found that ACTH and 17-hydroxycorticosterone levels were increased under stressful conditions only when the subjects felt helpless because they were unable to fit problem-solving into a pre-existing conceptual schema (Mason et al., 1976). Henry and Ely (1980) concluded that in the initial reaction uncertainty produces catecholamine-induced exploratory behavior. When the uncertainty is unrelieved, then a secondary diencephalic limbic activation produces prolonged tonic pituitary diencephalic response.

Stage III: Primary Process

The onset of kindling mandates new sensory and perceptual experience. The massive new sensory input and continuing high drive pressure, together with a hyperalert highly generalized identity consciousness promotes the formation of new identity structures, for the new experience comprises a basic new image of reality around which the drives coalesce. The highly charged sensous imagery and highly real experience therefore integrates itself in a hierarchically superior position in the long term memory system. New fantasies may now crystallize around these generic memories. The new identity structures form the basis for new identity constructs, with attendant new social organizations, while the generic memories and attached fantasies form the basis of a new regulation of the drives.

Stage IV: Primary Revision

Up to this point, the adaptive and growth processes run similar courses. Now, the growth process (also known as the creative process) with which we are specially concerned, diverges from the adaptive process. Using dominant hemisphere mechanisms, new action programs inspire new ways of acting. Indeed, a whole range of programs can emerge out of the collaboration between the new functional identity and the attendant organizing fantasies.

Stage V: Secondary Revision

Because they engender positive social responses, new behaviors yield pleasure, and so remodel the previously challenged identity. Prior identity and ego functions are superseded. They become a part of the structure of the newly revised ego functions. Thus ego functions acquire a hierarchical structure of their own. Neuropsychological systems now have access to a new distributed cortical territory. System-consciousness has achieved a new cognitive frame based on the altered identity construction, which allows the reception of previously novel experience without disruptions to the system-consciousness. In successful identity change the traumatic event is no longer traumatic. It acts, in fact, as a link from the previous to the new identity stage.

Schizophrenia as a Distortion of the Growth Process

Schizophrenia is a distortion of the process for psychic change. An adult process, it produces pathological identity change in response to insoluble problem-solving difficulties. (1) The onset of the process, (2) the regression, (3) the acute phase, (4) the subacute phase, and (5) the chronic phase correspond to the sequence of stages that occur during normal prolonged identity change.

Stage I: Problem Solving Failure in Schizophrenia (Onset)

Conventional wisdom has it that the ultimate prognosis is worse in schizophrenic illnesses that begin in adolescence without noticeable precipitating factors. My own belief is that a precipitating factor does exist in these cases. It is an endogenous one, and its arising inside subjective identity tendencies has made it obscure. The covert precipitating factor I mean is the schizophrenic's developmentally mandated experience of falling in love. What also makes this experience covert in some schizophrenics is the remoteness of the love object who may even not be known in any real sense by the schizophrenic person.

To put the case in developmental terms, object love appears to be a prerequisite of the entrance into the final stage of adolescence. To go on to adult life and to leave egocentrism and narcissism behind, the developing adolescent must have an experience of loving another individual in an object-related way. Loving in an object-related way means becoming able to libidinize (actuate) the real social world by means of granting a full subjective reality to another person who exists in that world. This is a prerequisite of sharing. This biologically mandated shift in libidinal (action) cathexis triggers an impossible identity challenge for the preschizophrenic person. Unable to complete the necessary developmental transaction of loving, the schizophrenic is destined continually to repeat this very process, without ever being able to culminate it.

The schizophrenic person's failure to form a successful love relationship produces a prolonged unremitting shame reaction. The "I" in the ego-synthesizing proposition "I am in love" is deprived of the subjective means to form an integrated subjective center of identity that is serviceable for adult life.

In his prior maturation the latently schizophrenic person has developed an intense devotion to narcissistic organization. This narcissistic organization expands greatly in the early to middle stages of adolescence, which are themselves typically narcissistic. In the last stage of adolescence, development mandates a limit to narcissistic expansion and a desire to fall in love. Falling in love challenges the organization of the latent schizophrenic's whole narcissistic world. The experience of losing the subject in favor of the object, which occurs as a transitory stage in the process of falling in love, terrifies the schizophrenic, who feels that his or her narcissistic mental integrity is being lost without any hope of restoration. During the onset of illness, the schizophrenic's falling in love holds all the terror of losing one's mind.

I believe that falling in love evokes a neurotransmitter outpouring of dopamine resources. The crisis of entry into adult life will require the schizophrenic individual, as much as any one else, to proceed on the basis of coherently conceived action plans. When schizophrenics pass into this stage they are unable finally to use the outpouring of action resources to synthesize coherent identity for themselves as loving persons, worthy of giving of their subjective resources.

The resulting sense of deep shame, of failed subjective identity strivings, evokes intense claustrophobic anxiety that inhibits any goal oriented activity. The inability to produce a stage-appropriate change in subjective identity and to synthesize overall identity as an adult produces a long regression-inducing panic. The outpouring of dopamine continues to compel action. Because there is no hope of consummation, the onset schizophrenic experiences subjective deregulation that is attended first by sleeplessness and finally by a full depletion of serotonin resources. This triggers the entire system-consciousness deregulation.

Luria (1980) reports on dramatically increased frontal activity recorded in the computer averaged EEG (CEEG) in cases of acute schizophrenia. This data confirms the impression that the system-consciousness is triggered into continuous activity in the acute phase of the illness:

> Similar results can be observed in patients with the paranoid form of schizophrenia, in whom a continuous state of stress, associated with static points of excitation, is also manifested as an increased number of synchronously working points in the frontal cortex. Administration of chlorpromazine, reducing this state of stress, was also shown to sharply reduce the large number of synchronously working points just described above in the frontal cortex. (p. 285)

The number of synchronously working points in the frontal area is a measure of the amount of system-consciousness work being carried out at a single point in time. In the example used to demonstrate this point, it appeared that a schizophrenic individual recruited almost his whole frontal cortex to solve an arithmetic problem. Fifteen minutes later, after taking his chlorpromazine, the frontal cortex was little involved in the work of solving such problems (p. 291).

The intense frontal recruitment appears to be initiated by left frontal failure at integration. Schefflin (1981) reviews the evidence for schizophrenics' difficulty in this area. He concludes:

> There seems to be little doubt that schizophrenic people have difficulty in left hemisphere integration and in the integration of left and right hemisphere interaction. My belief about the matter is this: If the schizophrenic cannot adequately integrate hemispheric functions, he might have difficulty in relating verbal and other specific task behaviors to the spatial and other contexts in which he is to act. (p. 128)

The combined evidence of dopamine disturbance and left hemisphere disturbance in schizophrenia suggests (within the framework of the structural paradigm) that schizophrenics suffer from an inability to integrate subjective identity, and from an inability to regulate subjectively initiated neuropsychological processes. Thus, in the schizophrenic onset disturbance, left hemisphere difficulty in problem-solving reaches an acute overwhelming intensity that triggers a system-consciousness breakdown of identity context. Thus the schizophrenic person's inevitable feeling at the onset of the acute phase is that subjective identity has been lost.

Stage II: Regression

The schizophrenic now enters into a phase of relinquishing synthesized aims and goals. Regressive, more primitive, earlier developed versions of development temporarily organize aims and goals. A feeling of panic engendered by the feeling of losing the center of subjective existence follows.

During the period of system-consciousness desynthesis, symptoms of deregulation appear in each channel of identity integration. Laing (1965) described the symptoms of subjective deregulation of identity feelings. He said that the acute schizophrenic person feels a sense of imploding or exploding subjectivity, that leads to feelings of subjective emptiness or of having turned to stone.

Self and object integers of system-consciousness identity also show symptoms of disintegration during desynthesis. Feelings of depersonalization and feelings of affective emptiness attend the disintegration of medial self structure. Feelings of *derealization,* of people being lifeless puppets, robots, or phonies all indicate disintegration and the initial fragmentation of the structures which organize social reality. The deneutralization of the nondominant hemisphere structures of identity liberate aggressive (reality) energy, which leads to desperate attempts to change reality conditions to reimpose some order. Cognitive rationales are developed to explain the changes in all of the bases for identity organization. These explanations begin to have a delusional quality in that they describe intrapsychic system-consciousness changes but no longer have validity as social statements. At this point the schizophrenic's statements begin to have a peculiar metaphorical ring, since the statements describe intrapsychic process. In the regressive phase disintegration of function follow the hierarchical release. As the ego and neuropsychological functions lose their dominant hemisphere regulation, other classical symptoms appear. Shimkunas (1978) describes this process.

> . . . The predominance of left-hemisphere activation in acute and responsive schizophrenics appears to reflect an extremely active problem-solving orientation. The intensity of this activity may produce high levels of information overload . . . (p. 212)

The acute informational overload and consequent intensified attempt at left hemisphere integration in acute schizophrenia deregulates the dominant hemisphere controlled speech system. Since the speech system fails to be of much use in providing context-matching, the schizophrenic uses it in an increasingly regressive way in an attempt to force more primitive ego-systems to provide missing satisfaction.

The Bleulerian description of *association defect* in schizophrenia also can be ascribed to the breakdown of frontal level identity components in regulating neuropsychological functions. The breakdown in associations is, I believe, a result of subjective, left hemisphere fragmentation of the conceptual basis for subjective integrity. As we have seen, the schizophrenic loses coherence of planning and goal oriented behavior as subjective integrity wanes. The result of imploded or exploded subjective identity is the deregulation of sequence in the propositional sequencing which derives from identity integrity in the subject.

Schefflin (1981) gives a good summary of the disorder of linear sequencing of speech as a communication function in schizophrenics. His description emphasizes the left hemisphere disruption in the orderly contextualization of speech.

> The patient uses strange or unusual ways to order the chronology of events. The sequence is not continuous, a problem we classically speak of as fragmentation. The patient is tangential, i.e., he takes off in one direction, then suddenly begins again and pursues another line of development. Also the schizophrenic person seems to have difficulty remembering where the story is going, what point he is trying to reach . . . (p. 99)

These problems in the linearity of narration are of course, as we learn from Luria, left hemisphere difficulties in the organization of planned behavior. In these difficulties in speech communication the lack of subjective integration expresses itself directly.

Schefflin describes the overactive initiatory muscular system in the acute phase:

> Those with acute psychotic disorganizations and paranoid trends are endlessly overactive. They flee and pace. They move from location to location or, when they are obliged to remain in a space or room, from posture to posture. At the very least they show a fidgeting restlessness of the hands and feet and a constantly shifting gaze. In these tense and active patients there is usually an extremely high state of muscular tonus . . . (p. 76)

Thus, Schefflin comments on the symptoms of acute action system and motor deregulation.

The desynthesis of system-consciousness leads to the disintegration of nondominant hemisphere based ego and neuropsychological functions as well. Many writers have commented on the sensory overload that begins to plague schizophrenics as the regression continues. Ingvar (1980) describes this phenomenon:

> Schizophrenia is accompanied by a change in the awareness of the outside world, as well as of the body itself. Trivial sensory signals—auditory, visual, and somasensory—interfere constantly with awareness, which leads to abnormal distractibility, interruption of the flow of thoughts, and confusion. Even proprioceptive stimuli which usually never reach consciousness may be perceived with abnormal intensity. (p. 120)

If we take the perceptual disintegration as evidence of lateral nondominant deregulation, we can take the emergence of diffuse body sensation and interference with affect integration as related to medial nondominant hemisphere processing difficulty. As acute schizophrenics lose their sense of synthetic identity integrity, they experience a diffuse affective response as if it did not emanate from the self. Thus, in the bewildering regression of mental function, schizophrenics find themselves crying without realizing that they are sad, find their hearts beating fast without knowing they are anxious; in fact with the identity functions deregulated they may feel that alien experiences are happening to a body that is not entirely theirs.

Stage III: Acute Schizophrenia (Primary Process)

If the schizophrenic process reaches this stage, kindling occurs. New structures for experience emerge out of renewed contact with long term memory. Heightened sensory and motor experience, and intensified discharge tendencies break through system-consciousness selective inhibition. The schizophrenic experiences waves of intense emotion that are shaped by wakened long term memories. At this stage the memory and fantasy waves overwhelm the current, already destabilized identity synthesis. When *significance experience* shaped by kindling occurs, the schizophrenic process has passed a point of no return. Heath (1977) has recorded high spindling subcortical EEG in the amygdala and hippocampus of schizophrenics. He interprets this activity as indicative of painful emotional experience. The abnormal, limbic nuclei provoked spindling indicates to me that long term memory of a generically painful kind is being provoked as the hippocampus facilitates the reception of new sensory experience. I theorize that this new sensory experience termed *significance experience,* together with the lack of resolution of all identity keyed programs, provokes neuropsychological level transformation forming new experiential structures. I call this the true acute stage because the neuropsychological changes are irremediable. This is the time when the past organization of mental processes takes over the present and overwhelms the existing structure of identity completely.

Stage IV: Subacute Schizophrenia (Primary Revision)

The return of earlier long term memory experience makes up the nucleus of new, pathologically delusional identity structures. Before this happens, fragments of new experience begin to coalesce around the earlier organizing fantasies of earlier periods. The formation of pathological structures is analogous to the formation of scar tissue. New permanent structures develop and distort a portion of experience.

We may understand the formation of (1) significance experience, (2) organizing delusions, and (3) organizing hallucinations within the framework of the paradigm. In the prodromal and regressive stages, delusions consisted of the experience of disintegrated identity functions, and hallucinations related to the inability to localize thoughts and communications from the outside. At this subacute stage of the syndrome, however, delusions and hallucinations are developed as restitutional structures. The deregulated speech, motor, and perceptual systems begin to coalesce around new fragments of experience that have incorporated material from earlier eras forming structured delusions and hallucinations that contain thoughts developed in a pathological core of thought formation. The lack of integration of identity functions still accounts for the form of the hallucinations and delusions, but the content is determined by the pathological changes in neuropsychological structure. As primary revision goes on, the delusional explanations of experience take on more and more explanatory value.

Clinically, we see that the hallucination during the first three stages of schizophrenia tend to consist of answers to subjectively vague questions. These questions relate to inchoate identity program struggles. Later, in the restitutional portions of the process, we may note clinically the hallucinations voice thinking which contains thought patterns and instructions that were formed much earlier in life.

Thus, schizophrenics weave their significance experiences and hallucinations into a restitutional pattern though which they achieve a pathological and delusional revision of their identity. As the pathological reintegration begins, neuropsychological functions (meaning, salience, percept formation, internalization of speech fragments) reorganize under the auspices of newly emerging nondominant hemisphere identity functions; for, while the dominant hemisphere triggered the desynthesizing process, the nondominant hemisphere restores order. Meaning and goal orientation now become subordinated to nondominant hemisphere organization. Characteristically, schizophrenics make an inspired declaration at this stage. "Now I know what this all means." They explain the relations in their world and phenomena of their illness through a delusional but thorough family romance mythology.

Freud was the first to realize that these secondary delusions of schiphrenia are restitutional, that they make up for the original inability successfully to achieve a love relationship through the formation of false object structure. Toward his conclusion to the Schreber case analysis in *Psycho-Analytic Notes on an Autobiographical Account of a Case of Paranoia,* Freud observes:

> . . .Schreber's 'rays of God', which are made up of a condensation of the sun's rays, of nerve fibers, and of spermatozoa, are in reality nothing else than a concrete representation and projection outwards of libidinal cathexis; and they thus lend his delusions a striking conformity with our theory. His belief that the world must come to an end because his ego was attracting all the rays to itself, his anxious concern at a later period, during the process

> of reconstruction, lest God should sever His ray-connection with him—these and many other details of Schreber's delusional structure sound almost like endopsychic perceptions of the processes whose existence I have assumed in these pages as the basis of our explanation of paranoia. (S.E. Vol. XII, pp. 78–79)

Freud's analysis of Schreber (through Schreber's autobiography) emphasizes the meaning of restitution in paranoid delusional illness. Through his delusions Schreber replaced Flechsig, the psychiatrist whose love he yearned for, with delusional rays of God which provided him with the missing love, and with the generative—I would say dominant hemisphere sources—which he felt had dried up in himself.

I believe that the various types of schizophrenia, *paranoid, catatonic,* and *hebephrenic* (disorganized), may be categorized on the basis of the neuropsychological type of restitution which is employed in the reintegration of the objective identity integers. The nature of the restitution has much to do with the original character structure and the mode of relating to other people. Thus, the paranoid incorporates new salience experience (significance) in the formation of new maladaptive restitutional object structure. A secondary system of delusions comes to comprise a false social structure that maintains the paranoid person's sense of social identity. This false structure explains the whole psychotic experience of having lost a sense of objective boundaries of identity. In fact the whole memorial past—old traumas—may be reformulated to conform to the delusional restitution. This reformulation takes the old organizing fantasies, with their primitive imagery, and elevates them to present identity themes. The family romance theme of the old fantasies is especially likely to find new organizing delusional significance.

Thus, the paranoid individual commonly reformulates identity along grandiose lines of the individual having mandates directly from God, and/or important parentalized social and political forces. The experience of primary delusion is also explained by way of these family romance themes as a materialization of inherent greatness. The greater the degree of romanticization and secondary delusional elaboration, the greater the difficulty the paranoid person has in adapting to present social reality.

I believe the catatonic form of schizophrenia may come to represent a nondominant hemisphere form of restitution in which postural and imitative motor themes are used to reorganize the object-self. The catatonic is drawn into the other person's posture. Drawn into the posture of their externalized, object-selves catatonics feel they can no longer find themselves, for their initiating motor organization is lost, and they feel they exist only through the feedback from their imitation of others. In their restitution catatonics feel that only the object has significance. Therefore they forgo any attempt to reconstitute their own subjective intentionality.

In the hebephrenic (disorganized) subtype, an even more primitive and more medial focus of identity organization is used as the basis of the restitutional process. The hebephrenic takes such neuropsychological components of self as affects, grimaces, language idioms—mere pieces of his or her thought core—and treats them as if they were themselves whole object structures. Such part-objects are used as externalized organizing object-selves, upon which the disorganized schizophrenic organizes a momentary sense of identity. It can be seen with each subtype that the shift of mind/brain processing proceeds to the nondominant hemisphere and to the posterior organization of neuropsychological qualities, rather than the anterior organization of extended identity functions.

Stage V: Chronicity (Secondary Revision)

In the growth process, as in the dream process, this is the stage at which the new forms of consciousness are organized and made to fit into the logical structure of conceptual identity. In the growth process, this is the stage at which the individual must test the new products of the inspirational earlier stage of structure formation to see if they truly lead to a resolution of the identity altering problem which began the growth cycle.

This is the stage of consolidation of the new maladaptive identity structure in the system-consciousness. The change in subjective identity thus produced is usually one in which motivation has been permanently reduced and in which previous goals have paled without being replaced by new ones. The original problem has been solved, however, because along with the reduction in consummatory expectations is a reduction in the original left hemisphere drive pressure, and a reduction in the attendant anxiety. The nondominant hemisphere identity structures attain new coherence in this phase of chronicity formation.

Whatever the subtype of schizophrenia, the new structures of experience are organized under the aegis of a new generalized identity structure in the nondominant hemisphere. In this phase, the lack of distinction between object-self and object becomes structuralized and permanent. The self loses its significance as the organizing component of objective identity in chronic schizophrenia. Affect is no longer a guiding communication, either intrapsychically or interpsychically. The object-self also loses its coherence, becoming a subcategory of the object. The person feels in chronic schizophrenia that official identity in the world, object-self identity, is subsumed under the notation of 'patient' or 'schizophrenic'. Some other person, or more often an actual institution, becomes the organizing object that maintains identity functions, subsuming the patient within that framework. This stage of chronic symbiotic affiliation with an institution may be termed *institutional transference.*

The more chronic schizophrenic individuals become, the more similar they appear to the outsider. This is not only due to the effects of medicine in rendering the faces relatively immobile, and the body relatively stiff and awkward, it is also due to the loss of subjective aspects of character and style, and to the similarity produced by the extreme generality of objective identity reached in the schizophrenic's system-consciousness.

I believe that the differences in character which distinguish even chronic types of schizophrenics have to do with the kinds of primitive organizing fantasy that find expression in the final object-self identity structures. To some degree these identity structures retain the characteristics of the individual's particular schizophrenic processes.

In weaving a continuous grandiose tapestry of his or her heroic origins and persecutions, the paranoid uses the obsessional attributes of the nondominant hemisphere to "prove" the validity of such claims. The catatonic reverts to the simple good/bad overgeneralized religious motifs that made him or her a model child. The hebephrenic uses the habitual massive denial that was a lifelong habit to produce an inane personality free of conflict. Some degree of restitutional object choice remains unassimilated in the final structure of identity that brings the schizophrenic process (as a process) to an end.

The end stage, the burnt-out phase, is reached to a variable degree by different individuals. In my view of the illness, there may be any number of acute episodes before the end stage is reached. Each acute episode follows the same sequential process and results in further breaking down of subjective structure and some further development of a generalized object-self which structures the chronic schizophrenic's experience. Thus, depending on the ultimate degree of revision, the schizophrenic person has different degrees of remodelling of the intrapsychic structure according to his or her own proximity to the end stage structure of the illness. I have hypothesized that the formation of the *generalized object-self* is in actuality a reaching forward to the middle age crisis remodelling of the personality. Thus, the middle age schizophrenic often experiences a much lesser degree of intensity of symptoms. The dopamine pressure often abates after the middle thirties, and the impetus for continued cycles of schizophrenic process drops off. In middle age, the schizophrenic person often becomes a rather tame, empty individual caricaturing a certain kind of impoverished adaptation in middle age undertaken by some individuals.

The truly final stage of generalization of the object-self may be marked by stigmata of tardive dyskinesia. Years of taking dopamine blocking agents supersensitizes receptor sites, and destroys many of them, producing movement disorder. Tardive dyskinesia also undoubtedly affects the hierarchical organization of subjective functions, producing a kind of further paralysis of voluntary speech and planning. Thus, the overall effect of the illness—and its cure—is to simplify frontal identity structures to the point that they operate simply, and as infrequently as possible. Problems are not taken on. The schizophrenic in chronicity expects the institution (the externalized generalized object-self) to make decisions.

Under these circumstances the neuropsychological focus of conscious processes shifts posteriorly away from the frontal organization of the system-consciousness. It also shifts to the nondominant object hemisphere. There is very little organized subjective structure at an identity level operating in neuropsychological problem-solving. When a chronic schizophrenic is forced to make life decisions, he or she often suffers a regression and a shorter or longer episode of schizophrenic process in response to the challenge. Thus, the degree of right hemisphere posteriorward shift in the focus of cortical operations should be a gauge of the degree of chronicity the patient has reached.

Ingvar (1980) reports on this phenomenon:

> In normal human subjects there is, at rest, a typical "hyperfrontal" activity distribution. The activity is considerably higher in frontal parts of the cortex than in parietal and temporal parts. This pattern is often less conspicuous in chronic schizophrenics. In fact, the pattern may even be reversed in such patients.
>
> We have also found that the more inactive, mute, and autistic the patients were, the lower was the flow in the frontal regions of the brain. The more evidence there was of cognitive disturbance, the higher was the flow in parietal and temporal regions of their brains. (p. 107)

The discussants noted that these blood flow studies correlate exactly with information derived both from computerized EEG studies and from glucose utilization studies, both indicating decreased frontal usage in chronic schizophrenia.

Shimkunas (1978) reviews literature relevant to the dominance shift in acute to chronic schizophrenia. He concludes:

> . . . The acute stage may be characterized by an overuse of the left hemisphere in attempts to problem solve, leading to an ultimate shutdown of central attentional processing of external stimuli . . . The chronic stage may then represent a shift from the inadequately functioning left hemisphere to predominant right-hemisphere activity. (p. 213)

Shimkunas ties this shift to the conceptual generalization that occurs in chronic schizophrenia. He believes that autism is due to the fact that the schizophrenic uses high level phrases without reference to concrete experience, long term memory, or problem solving. Shimkunas points out that the nondominant hemisphere organizes perceptual data, including images from the real world. He concludes that in the process of overgeneralization new higher level abstract categories must be made. Thus he observes:

> Autistic and otherwise incomprehensible statements that are so common among schizophrenics thus become understandable as attempts to combine different levels of abstraction to enable the individual to sustain a comprehensive grasp of his or her stimulus inundated external reality. (p. 222)

We may see the overgeneralization process in the nondominant hemisphere as a means in the acute stage to maintain the ability to integrate an increased sensory load, but in chronicity as a simplified intrapsychic control and regulation mechanism. The broadened categories of generalization draw on the frontal capacity to form conceptual hierarchies. The overcomplete conceptual categories allow for the formation of some intrapsychic order and some reflective stability.

Chapter 4

THE PROCESS OF ADAPTATION AND THE MALADAPTIVE AFFECTIVE SYNDROME

Freud described depression (melancholia), (*Mourning and Melancholia,* Vol. XIV) as pathological identification with a lost object. Underlying the unconscious ambivalence that Freud cited as the reason for pathological identification is a two year old's failure to resolve the separation-individuation conflict of infancy. Depressive individuals fail to distinguish completely between their own selves in the world (object-selves) and the necessary object who mediates their participation in reality by educating them about the world. Possible reasons for this inability are (1) parental possessiveness, (2) restrictive parental competitiveness (which creates dependency), and (3) a lack of an educating figure (which creates a deficit in experience). For any of these reasons, and for biological reasons as well, the person who is liable to depression develops both an overly intense need for a mediating object and a resentment against the experienced inadequate mediating figure.

Under these conditions, the future depressive spends a major portion of identity formation work identifying with the object's traits.

When the object dies or becomes unavailable, and if there is no substitute object, pathological and maladaptive identifications with the lost object occur. The massive input of external features of reality precipitated through the loss of the mediating object causes the depression prone individual to erect the lost object inside the object-self as a mediating identity structure—causes the person, that is, to replace the lost external object with an internal one. The erection of this resented, ego-alien structure inside the object-self creates a precondition for an attack on the resented portion of the object-self.

The present chapter translates Freud's insights into a description of the major affective syndromes, mania and depression. The translation focuses on defects in identification, the nondominant hemisphere process that relates the medial self to the lateral object-self functions of identity. We will discuss mania first, since mania is not only the first step in bipolar disorder, but also seems to be a brief but important phase in the onset of any major depression. Interestingly, the manic portion of bipolar disorder and the short hard manic premonition of major depression both pass through the first three stages of a pathological version of one of the processes for normal identity change, the adaptive process. Major depression represents a complete maladaptation. Thus, I have treated major depression and its precursors as parts of a maladaptive syndrome.

STAGES I AND II: THE ADAPTIVE PROCESS

The nondominant hemisphere triggers the adaptive process. The first two stages (problem solving and long term memory invocation) in this process are familiar to us as the initial stages in the stress response. They are also familar because they are the nondominant hemisphere's method of problem solving. These two stages encompass the response to novelty that cannot be dealt with within the cognitive framework that is available through the ordinary exercise of objective identity function.

Averill (1979) discusses these two stages:

> ... Events are meaningful only to the extent they can be assimilated into some existing cognitive model or structure. If they cannot be assimilated, then the relevant cognitive structures must be altered to accommodate the environmental input. What this means in practical terms in that the regulation of stress may require a two-stage process. The first stage involves the creation

> of an appropriate frame of reference, and the second stage is the presentation of specific information or instructions with regard to the nature of the threat, possible coping responses, etc. (p. 383)

The adaptation thus consists of a search for already existing cognitive structures, and if these are not present, then for a modification of existing conceptualizations incorporating some of the novel bits of information. Rose (1980) discusses these adaptations, including the hormonal responses which modulate them.

> We have become impressed with the fact that individuals adapt or accommodate to most stressful stimuli upon being re-exposed to them, and thus the stimuli lose their stressful quality. What we are saying is that most stressful psychological stimuli that have been clearly shown to be provocative of a variety of endocrine responses are acute or novel in nature, and that upon re-exposure, endocrine responses extinguish reflecting adaptation. (p. 255)

In adult life normal adaptation stops at this point.

In the course of development, adaptation proceeds through three further stages. Stage III involves primary process kindling with identification. This provides an influx of new experience and new features of reality. Stage IV, primary revision, begins the revision of the object and object-self representations. The revision is based on an assimilation of the new experience and the new features of reality. Stage V, secondary revision, completes the remodelling of the objective identity functions. This revision includes a reconceptualization of the object-self including its social position. Taken together, these five stages of the adaptive process comprise the formation of character change as it is accomplished in the nondominant hemisphere. During the developmental process there are periods in which nondominant hemisphere change triggers the change in system-consciousness. As we saw in Chapter 1, these alternate with periods of dominant hemisphere initiated growth. These periods of nondominant hemisphere triggered adaptation to reality include the anal period, the latency, and early adult life.

The natural decline in norepinephrine availability in midlife makes this a period of adaptive change and reorganization of long term memorial processes. This change in system-consciousness that is mandated by the life cycle biological events can also give rise to the depressive syndrome that is so common in this era. The early stages of the process of adaptation exhaust norepinephrine and serotonin supplies and along with this comes a supersensitivity of beta adrenergic receptors and finally a destruction of receptors. It is during stage three of stress exhaustion that kindling sets in.

According to Rosen (1984):

> Preliminary data indicate a decrease of brain noradrenaline levels during the kindling process as well as a reduction in the number of beta-adrenergic receptor sites. Depletion of forebrain noradrenaline markedly accelerates the development of amygdala kindling. Changes of neurotransmitter metabolism during the kindling process might have important behavioral consequences . . . The most remarkable change is a reduction of rapid eye movement (REM) sleep after the kindling has been established. (pp. 29–30)

Thus we have evidence that kindling is a process that accompanies those noradrenergic changes that we associate with the onset of major depression. What is more, the early latency of REM sleep in major depression has been taken as a virtual biological marker of depression.

The Process of Manic Induction

In cases that show a disposition toward affective illness the initial stages of the adaptation process—(1) problem solving, and (2) novelty anxiety that provokes long term memory solutions—can fail to produce the desired adaptive response. When this happens the patient progresses to a third stage of adaptation. This acute stage involves primary process thinking induced by limbic system kindling which I theorize provokes intense indentification. This acute stage corresponds to the induction of mania.

With the onset of mania, the individual switches from lateral nondominant hemisphere neuropsychological functions to medial ones. During this dramatic switch the manic feels the imminence of gratification. In mania, the whole object world seems a fruit ripe for the picking. The manic feels that he or she possesses the whole world. Sometimes a blank dream accompanied by orgasm (and probably by the most intense norepinephrine enforcement of REM dreaming) launches the phenomena that accompany the manic illusion.

We can make certain deductions from the phenomenology of acute mania. The manic process recapitulates the preparation for and passage into a state of discharge and satisfaction. Psychologically this state consists of the feeling that the object world has yielded conditions of a satisfactory match. This is to say that goals and means are simplified, identity is reduced to its essentials and the neuropsychological discharge process is laid bare. In mania the signal for nondominant hemisphere discharge release is given over and over again. The result of the false signal is intense mental pain, the most intense frustration, an immediate tide of rising anger, and an attempt to force reality to yield the satisfying conclusion which is being withheld. Waves of such identity-signalled, limbic-facilitated, hypothalamic-relayed, hormonally-fixed and modulated activity cycles punctuate the manic course. Each novel stimulus taken as an opportunity for imminent gratification becomes an acute stressor instead. Each novel stimulus becomes the missing link, the key to resolution of the lifelong problem of failed gratification.

In sum, the manic onset brings the world into immediate conjunction with the affective and bodily self. The world is felt to be taken in, swallowed, and reassembled inside the self as a permanent source of gratification. To the extent that this identification can lead to altered behavior, there is indeed the possibility of producing the means of satisfaction. A result is that manics act as social magnets, drawing other people into their orbs with promise of new identity, experience, and complete satisfaction. To put the case in neurological terms, content in the outer world of reality is generalized and represented by single novel signifiers: the medial sphere of the self expands so that each aspect of the self is exaggerated. It is accurate to use psychoanalytic terminology to refer to this process as a massive *denial* of and replacement of the world via the identification-mechanism.

Research findings bolster conclusions based on direct observation of manic stages. Neurotransmitter findings, for example, indicate that there is a brief increase in norepinephrine in the acute stages of mania that coincide with manic induction. That is followed by rapid decreases in norepinephrine and serotonin levels thereafter. According to Sweeney and Maas (1980) who reviewed the literature prior to 1980:

> . . . it is now well documented that when bipolar patients (that is, those patients who have periods of both mania and depression) are studied longitudinally, urinary MHPG is significantly higher during mania than during depression . . . (p. 165)

MHPG is a breakdown product of norepinephrine and is commonly used as a gross assessment of the ongoing utilization of norepinephrine. Thus, neurotransmitter evidence shows that manic induction begins with a phase of high, but poorly integrated reality contact. Ingvar (1980) finds markedly increased frontal blood flow in the five manic patients he investigated (p. 129). This is compatible with the conclusions reached on the basis of neurotransmitter evidence.

Buchsbaum (1980) has investigated the characteristics of bipolar manic-depressive patients on evoked potential stimulus. Some of his summarized findings are relevant here.

> ... The bipolar patient, relatively insensitive at very low stimulus levels but with large EPs for intense stimuli, appears to have a high setting on his sensoristat ... (p. 237)

Buchsbaum's conclusion is used in support of his contention that manic depressives are stimulus seekers, who seek novel stimuli, constantly transforming them as if they were crucial pieces in the gratification puzzle they are attempting to solve at all times. Buchsbaum continues:

> ... The bipolar patient's diminution of EP amplitude on repeated testing and reduced recovery cycle may be examples of defective balancing of stimulatory need and inhibitory processes. Pain tolerance may be another pathological example of stimulatory need and disturbed stimulus-intensity modulation. (p. 237)

In my view the high pain threshold and reduced recovery threshold have to do with the manic's overinvolvement with medial processing, tolerating pain in order to go ahead with the gratification process. The manic periods correspond to periods when increased norepinephrine expenditure is possible and when stimuli are processed at a high rate. These periods of high stress give rise to neurophysiological responses, including increased TSH output, increased CRF–ACTH output, and increased endogenous opioid moieties that block the incoming pain system. During intercurrent periods of exhausted resources the bipolar patient is a stimulus seeker, unable to process reality sufficiently well to solve the gratification problem.

Wolpert (1980) combines the psychoanalyst's view of manic induction with an experimenter's understanding of activated imminence in which patients experience hyperacute sensation and unusually vivid dreams that represent the overwhelming of the higher mental integrations.

> Clinicians have named this state ... "prodromal symptom state", "impending mania", or "hypomanic alert." Experimenters ...

> working with the same phenomenon . . . have called it more simply the 'switch' . . . Both the clinicians and the experimenters have noted the switch to occur more often at night, perhaps during dreaming. (p. 445)

Because norepinephrine is considered to be the activating agent in REM dreams it seems likely that a state of medial nondominant hemisphere activation would begin onset in sleep activation of the medial norepinephrine drive system. The vivid character of the sensation undergone in acute mania attests to the activated veiled state which occurs in extreme medialized experience, for in all stages of life in which new central organizing experience is being taken into the center of the core of thought formation, imagery is dressed in great sensuality. Presumably, vividness is an accomplishment of the kindling during new organizing experience.

Acute mania, then, corresponds to a period of activated identification process in which characteristics of the world are taken into new perceptual organization, forming new psychic structure. One thinks also of the inspirational stage of creativity in which all sensory experiences are heightened and felt to participate in the formation of new significance. The introduction of such a magnitude of new experience during acute mania deregulates existing structures at the ego and neuropsychological level for processing salience.

Deregulation of Ego and Neuropsychological Functions in Mania

The hyperactivity, pressure of speech, flight of ideas, and inflated self-esteem are all rather direct manifestations of an increased reality drive pressure consisting of noradrenergic stimulation beyond the bounds of serotonin system and gabaminergic system capacity to regulate. The manic feels that he possesses reality, that every aspect of reality is salient and in his or her service. The high level of processing exhausting the serotonin system also overwhelms the capacity to sleep. When sleep does come on it is full of quick intense REM dreaming. In fact, we might see the blank dream or orgasm as announcing the onset of kindling. The overwhelming of upper level serotonin regulation produces a destructuralization of identity functions of the nondominant hemisphere. Thus superego function becomes poor or absent. Involvement loses it moral regulation and reverts to simple pleasure-pain determinants.

On these bases I hypothesize that mania cannot proceed without ample supplies of norepinephrine, and that mania must come to the end of a cycle with the exhaustion of available supplies of new norepinephrine to infuse the system. One aspect of my norepinephrine hypothesis concerning mania is that the switch from lateral to medial process simplifies the object world, producing the illusion of plenipotential features which produces an economy of norepinephrine. The decreased lateral binding because of vast *stimulus-generalization*, allows for increased central and affective processing of norepinephrine. I believe that the decreased lateral binding is interpreted by the system-consciousness as a signal of imminent gratification, and that this acts as a *switch-point* to medial processing. Thus, in this hypothesis, the "break" with reality constitutes the precipitant for a cascade of medialization in nondominant hemisphere neuropsychological processing.

The amount of norepinephrine required to induce such a cascade effect requires certain neurophysiological preconditions. This is the reason, I believe, why mania tends to occur in the twenties when norepinephrine maximalizes for the problem-solving stability which is so vital to that stage of life. It is also a clinical statistical fact that mania decreases markedly in the forties and fifties. In my estimation this is because of the decrease in norepinephrine that accompanies the aging process.

The Process of Endogenous Depression

Endogenous depression often begins in the forties or fifties. Norepinephrine depletion becomes a biological problem in the midlife period and this mandates structural changes in system-consciousness. The problem in endogenous depression is a depletion of sufficient neurotransmitter to continue the kind of information processing that has carried the person, with some success, into middle age.

The person with a predisposition to psychotic or endogenous depression in middle age develops an unusual allegiance to structures of social authority. Seeing figures of social authority as possessing or granting social satisfaction, this 'field dependent' person heavily invests his or her norepinephrine in a mediating object who has, the person believes, access to authority and to the satisfaction it represents. This investment is a later life manifestation of the identification undertaken by the child who must face parental overpossessiveness, overcompetitiveness, or lack of mediation.

The characterological aim of the depressive is therefore to discover how best to manipulate the mediating objects and how best to manipulate society itself in order to achieve security and satisfaction. Personal productivity is not important in this scenario. Thus, there is a characterological distortion involved in the predisposition to depression. The person's own personality and values within a social context are not felt to be important. There is, so to speak, a moral defect in the lateral objectification of object-self. Instead of emphasizing the person himself or herself as a social integer, the depressive person feels that only other people have access and standing in society. The effect of this deformation of character is to make self-esteem dependent on social re-enforcement rather than on personal ability to identify important issues.

Now, as I have surmised, in middle age the concept of social structure undergoes a change. I have theorized that this change takes the form of understanding that authority is *vested* in social structure, that it is not there according to some inherent right. The withdrawal of some social cathexis, and the understanding that social power is mutable changes the middle-aged person's relation to authority. Eventually the middle-aged person develops a more generalized conception of the outside world, including his or her place in it. In this, society is vested with power only by individuals who willingly give that power. At the same time, in preparation for the closing phases of life, the aging person begins to see society as functioning within a more generalized context of life and God's existence. These global generalizations of the social order and the outer world of reality simplify the work of the nondominant hemisphere, which yields a greater economy of norepinephrine consumption.

I contend that wisdom, with its neurotransmitter economy, is beyond the scope of the depressive's thinking. The aging depressive person feels that he or she has been cheated by life, has given service to a false notion, has been deceived.This is the time of life when the depressive person concludes that far from being ideal, his or her mate is all but worthless. Physical disability and depression in the partner contribute to the devaluation process to the point that the partner is no longer viewed as a mediating object. Thus, the course of the life cycle challenges the ambivalent investment in the externalized object-self and challenges the very premises upon which the depressive position was maintained.

With the relative flagging of norepinephrine and dopamine resources, the depressed person feels more and more cheated and more prone to enter into major illusionary reconstructions of his or her life. These illusionary reconstructions may produce brief manic episodes or manic tendencies. They certainly produce prolonged moods of either a depressive or a hypomanic nature. These mood states may be seen as the anlage of the changes which occur in the course of the depressive process. The deregulation of the investment in the value of the externalized object-self projected into the mate or other vested persons results in the imminence of object loss.

The major form of anxiety suffered by depressives is the anxiety of failing to be loved. Thus, rather than experiencing the dominant hemisphere neurotic tendencies of claustrophobia or paralysis of intentions, the depressive often experiences the panic of agoraphobia, or fear of being alone or abandoned, for the major conflicts in the depressive are nondominant hemisphere conflicts centering on the disposition of anger. The depressive is afraid of exhibiting hostile feelings and therefore attempts to suppress such feelings and to give the other person the benefit of every doubt.

Sweeney and Maas (1979) conducted a clinical study of 24 depressed women, 16 of whom had the diagnosis of unipolar depression. They found evidence for a correlation of "stress anxiety" and increased norepinephrine metabolism as measured by urinary MHPG (p. 172). This data fits the hypothesis that depressive individuals increase the level of neuropsychological processing of the world in response to stress and in response to anxiety. A different kind of data confirmatory to this hypothesis is reported by Gershon and Buchsbaum (1979) who found higher AER on photic stimulation in the right hemisphere of affectively psychotic individuals. This tends to support the contention that depressively disposed individuals favor nondominant hemisphere processing.

Thus, we may conclude that the predisposition for depression is based on (1) an excessive use of norepinephrine in nondominant hemisphere processing of information about the real world, and (2) on a greater than ordinary reliance on context formation in the nondominant hemisphere frontal areas. This requires a great deal of serotonin system co-modulation to produce the selective inhibition necessary for increased processing. The search for salience is intensified in depressives.

The evidence converges, then, that endogenous depression is a product of norepinephrine decrement, serotonin depletion, and later hypersensivity of these co-modulators.

Stage III: Affective Psychosis Proper

As we have seen, when the initial stages of adaptation fail, when cognitive sanctity cannot be reimposed to deal with unremitting stress, then a kindling process supervenes that sets the identification mechanism into motion. This is the period of desperately calling other individuals in the attempt to recruit someone to fill the mediating object void. Failure to complete the identification process leads to the outbreak of frank psychotic symptoms of depression.

The kindling, bringing with it the invocation of all past mediating objects, stimulates a profound dreaming process in which the mother or her surrogates (ghosts of the past) appear in dreams and are hard to dispel in waking reality. Characteristically, at this phase of the depressive syndrome the patient is given to intense visual imaging, even visual hallucinating, of the lost object. At the same time as objective identity functions fail to structure the object world, auditory hallucinations also come into prominence. Characteristically, the psychotically depressed person now hears his or her name called out. The voice frequently emerges at moments of intense agoraphobic anxiety. The wish for the mediating object is so strong that any sound becomes a possible voice calling the person in some familiar way. The depressive psychotic begins to associate the voice of self-identification with the dream figures left over from the disturbed sleep and dreams of past keepers.

It can be easily noted that the mood congruent hallucinations of the depressive psychotic are full of superego fragments. The fragments of object-self and object representation clearly show the decomposition of the lateral identity functions. The disparaging comments of the hallucinations, and the delusional exaggeration of disparaged self-identification show the attempt at self identification through identification with a harsh social order.

Stage IV: Endogenous Depression Proper

Endogenous depression begins with failure of identification to restore cognitive order. The loss of the mediating object or the complete disillusionment with previously held mediating objects in Stage

III produced a period of prolonged psychic pain that set limbic kindling processes into motion again. The uneven reception of new sensory experiences in an already exhausted and supersensitive nondominant hemisphere (due to its norepinephrine and serotonin deficits) pushes the individual into maladaptation. Stage IV of the adaptation process becomes pathological with the deregulation of nondominant hemisphere identity, ego, neuropsychological, and neurophysiological functions.

Depressive psychosis is, in my view, not distinct from endogenous depression. Indeed, I believe that even without the overt hallucinations and delusions of affective psychosis, when the disturbance of functions reaches ego and neuropsychological levels, this is sufficient evidence for diagnosing psychotic process. Maladaptation begins when the affective self's resources are depleted, and no mediating object-self equivalent person can be found. The depressive then concludes that life is intolerable, that it is inevitably and inexorably unbearable. Loss of belief in the mediating object leads to a retreat into the neurotransmitter impoverished medial self. The world seems empty, meaningless, and increasingly distant. The locus of objective identity function shifts to medial identity function. The 'me', the affective self, is experienced as emptied of esteem.

According to the structured paradigm, affect realizations are system-consciousness assessments of the nature of norepinephrine flow in the nondominant system of regulation. Hence frustration affects are due to an overflow of norepinephrine energy that cannot produce the cognitive match required for satisfaction release. The affect of frustration induces a judgment that the neuropsychological work of the nondominant hemisphere has been fruitless. As frustration energy is depleted in Stage IV of the maladaptive process, the hopelessness and futility of not having sufficient energy available to mobilize the self to perform neuropsychological work is experienced and assessed.

The intense manifestations of guilt that also accompany endogenous depression indicate a process of system-consciousness inhibition of the medial self function. Guilt demeans the medial self and decries its inability to function in the world. It is therefore an insurance that the self will not undertake any energetic escapades in the world. In this appraisal, guilt is seen as a hierarchical signal to inhibit all nondominant hemisphere higher level functioning. In this context guilt is seen as a general mechanism by which system-consciousness conserves energy in a period of chronic stress. Thus, withdrawal, lowered self-esteem, and guilt are all seen as conservative mechanisms undertaken in extremity when identification mechanisms have failed to maintain intrapsychic coherence.

In Stage IV, then, the nondominant hemisphere ego functions are compromised by the deregulation of the hierarchically superior identity functions. Ego function impairment materializes, for instance, in distorted body sense. In the paradigm model the medial nondominant channel regulates the body consciousness as an entity in space. In severe endogenous depression the person often feels as if the body is rotting. This feeling must be determined by a lack of cathexis for the body as an object in space as well as by an awareness of the loss of autonomic tone that occurs in severe endogenous depression. The feeling of body rot is then an accurate metaphorical description of the erosion of the medial experience of body tone.

The lack of ability to concentrate, which is another major symptom of endogenous depression, may also be seen as a direct manifestation of the decrement of norepinephrine energy in the nondominant medial system ego functions. The ability to concentrate means that the serotonin system is inhibiting all but selected areas of nondominant hemisphere cortical distribution. It requires a vast suspension of data coming in from the world and a pursuit of particular data in great detail. Thus, concentration requires the ability of system-consciousness to utilize selectively the lateral nondominant channel. In endogenous depression, no energy is available for attending to planful relations to reality, for in psychotic depression, I surmise, the serotonin system fails. The exhaustion of the serotonin system must contribute to the early morning awakening as well as to sleeplessness.

The terminal insomnia of endogenous depression may then be taken as a disturbance at the neuropsychological level. Other neuropsychological disturbances may be understood as a result of dysfunction in the medial nondominant channel of regulation. The sighs, moans, and lamentations of depression are the affective contribution to the speech system, stripped of content. Laying bare the underlying energy rhythm of the nondominant hemisphere motor-system, the depressive's agitation is the sign of the inability to know where to place one's limbs and fingers in the world. Contrast this with the manic's rhythm-oriented dancing, still manifest in Stage III. When the bipolar patient reaches a stage of depression comparable to Stage IV, Depue and Monroe (1979) report that they show particular inactivity and retardation. Thus, it seems likely that bipolar patients go even farther in the exhaustion of their neurotransmitter resources than unipolar patients, and as a consequence they suffer further decline in their neuropsychological functions.

The world seems particularly empty, cold, and deserted to the depressive as he or she awakens in the early morning as some small spurt of norepinephrine awakens the self suddenly, to the remembering of total frustration which led to exhaustion and emptiness. The experience is almost like a dream of morning which fails to renew life. After a sharp anxious awakening, the depressive lapses into hopelessness. Under these conditions neurophysiological functions soon begin their own deregulation.

The autonomic signs of severe depression—the loss of appetite, loss of dreaming, loss of sexual potency, constipation, loss of body tone—must all relate to the change in the hypothalamic regulation of neurotransmitters and especially neuromodulators. The loss of identity function—the hopelessness and helplessness—presages a release of autonomic processes from frontal regulation. The withdrawal of neurotransmitter pressure, the destruction of noradrenergic and serotonin receptor sites, must be just one of the measures that conserves vital resources in the resistance stage of chronic stress.

Stage V: Chronic Psychotic Depression

In psychotic depression, the restitution of the object world must include a cognitive rationale for the breakdown of ego and neuropsychological functions. A new simplified identity structure must be developed. Self-identification becomes a prime motive of the restitutional phase. The quality of self-identification relates to the mood congruent nature of many of the hallucinations and delusions found in psychotic depressives. Just as the schizophrenic forms restitutional structures out of the experience of the psychotic process, so also do depressive psychotics form restitutional structures out of the experience of their psychosis.

The end stage of endogenous depression appears to be one in which the object-self is identified with a hostile, empty world. If this does not lead to death or suicide, the intrapsychic stasis this identification produces may lay the ground for an adaptive revival after some conservation of neurotransmitter resources has taken place, and probably after some help has emerged for the person who is in such chronically desperate straits. We know that the process of endogenous depression is often self-limiting; presumably, the long reduced psychic economy can lead to a revival of neurotransmitter resources in 6 months to a year.

Even after a prolonged period of time, in order for recovery to occur, there must be some change in the external world, perhaps some source of concrete help so that the object-self can be reidentified with the external world as a source of help. The development process of the midlife crisis also works toward such a resolution. It appears to me, moreover, that the resolution of endogenous depression may proceed through the mechanisms that Freud mentions in his paper on *Mourning and Melancholia*. Some self-knowledge capable of providing a cognitive conceptual framework may emerge in the conclusion that one is indeed selfish and worthless. With the passage of sufficient time, with the appearance of some helpful mediating person or institution, or with the renewal of neurotransmitter resources with medication or shock treatment, or all three, the depressive may come out of the endogenous process with a change in character structure in the direction of having developed an ironic or cynical attitude about the value of anything in the real world including oneself. We see some individuals who undergo repeated endogenous processes, each episode ending with a characterological change in the direction of an increased rationalized disillusionment with the world. This form of wisdom provides one kind of foundation for a midlife point of view that is economical in the sense that it is complete, simple, and generalized. Such a generalization of identity can help some individuals conserve their neurotransmitter resources and continue to function with a mitigated need for love and mediation in the world.

Chapter 5

PSYCHOANALYTIC THEORY

TOWARD A COMPREHENSIVE STRUCTURAL THEORY

Psychoanalysis has moved away from Sigmund Freud's brain-centered ideas about the structure of the mental apparatus. Two major writers in this area who resist the idea of localization in the brain, Heinz Kohut and Otto Kernberg, have built structural theories that depart from the original psychoanalytic impulse that *The Project for a Scientific Psychology* epitomizes; they have also contributed important insights about identity functions to psychoanalytic literature. The present chapter seeks to harmonize the new *self-theory* with Freud's biological view of mental life by placing both approaches inside the framework of the structural paradigm. It illustrates the relation of psychodynamics to the life and development of the brain.

Never allowing any part of his theory to controvert his fundamental belief in the unity of brain and mind, Freud labored all his life to find the biological foundations of the mental life he described in the psychoanalytic theory. His introspection based itself in a belief that at its basic levels mental life ceases to impress itself on us as psychology as its operation descends into brain structure. When he revised psychoanalytic structural theory in *The Ego and the Id* (1923), he raised mind/brain continuity to central importance. The new concept of *ego* was a vision of a system of structural functions that mediate between mind and brain while it synthesizes all aspects of identity.

Freud never gave up his pursuit of the brain/mind confluence. In his 1925 paper on *Negation*, he returned to his basic descriptions of the mental apparatus as he had first conceived it in *The Project For a Scientific Psychology*. This re-elaboration emphasized the identity functions inherent in the ego's synthetic and integrating capacities. Judgement is Freud's key synthetic concept.

> The function of judgement is concerned in the main with two sorts of decisions. It affirms or disaffirms the possession by a thing of a particular attribute; and it asserts or disputes that a presentation has an existence in reality. (S.E. Vol XIX, 1925, p. 236)

This is to say that Freud's *judgement* harmonizes the essential drive-quenching qualities as they are perceived in the two hemispheres, for to put the case in paradigm terms, Freud's judgement's ability to perform the synthesis consciousness calls each component of identity into play.

> . . . Expressed in the language of the oldest—the oral—instinctual impulses, the judgement is: 'I should like to eat this', or 'I should like to spit it out'; and, put more generally: 'I should like to take this into myself and to keep that out.' . . . What is bad, what is alien to the ego and what is external are to begin with, identical. (p. 237)

Thus, in the earliest ego (the earliest coalescence of the medial prefrontal functions at about three months), the *subject function* (the 'I' with its consummatory wish) and the *self function* (the 'me' with its image of being inside the body) fuse and form the medial core of the ego. The external or reality attribute (*object function*) is given a separate identity function as *it*. 'Is it good?' is the synthesizing question that determines the fusion of the medial components. 'Is it real?' determines the synthesis of the lateral components.

Freud (1925) says that judgement develops its second or lateral function as a step forward in development.

> . . . The other sort of decision made by the function of judgement—as to the real existence of something of which there is a presentation (reality testing)—is a concern of the definitive reality-ego, which develops out of the initial pleasure-ego. It is now no longer a question of whether what has been perceived (a thing) shall be taken into the ego or not, but of whether something which is in the ego as a presentation can be rediscovered in perception (reality) as well. (p. 237)

As a step forward in neurological development, the coalescence of the reality ego (at five or six months) completes the zonal differentiation of the system-consciousness through the integration of lateral zones that can selectively inhibit the limbic memory system to distinguish a present reality. As the reality-ego Freud's (1925) judgement performs another lateral function that is inherent to its synthetic role:

> Another capacity of the power of thinking offers a further contribution to the differentiation between what is subjective and what is objective . . .
>
> Judging is the intellectual action which decides the choice of motor action, which puts an end to the postponement due to thought and which leads over from thinking to acting. This postponement due to thought . . . is to be regarded as an experimental action, a motor palpating with small expenditure of discharge. Let us consider where the ego has used a similar kind of palpating before . . . It happened at the sensory end of the mental apparatus, in connection with sense perceptions. (p. 238)

Motor palpation, then, is the lateral integrating *agent function* that inhibits limbic level organized action sequences in favor of conditioned sequences that judgement renders efficacious.

In addition to making a distinction in the functions of judgement that is analogous to the vertical split between two medial functions (subject/non-subject) and two lateral functions (motor-palpating/perceptual-palpating) Freud introduces a distinction that is analogous to the horizontal split in the brain/mind between identity functions and ego functions. Freud's distinction is between *judgement* at a higher level and *thinking* at a lower level. I theorize that this split widens during epigenesis until consciousness truly develops a *reflective* and an *experiential* level of horizontal organization.

In the theory of 1923, the ego itself organizes consciousness. This idea of the ego replaces Freud's original system-consciousness in his topographic model and extends that system to include both what I have been calling identity functions and what I have designated ego functions. The creation of an ego that modulates self-concerns undoubtedly leads to a way of writing that necessarily anthropomorphizes the structural agency of consciousness. Indeed, we may ask if we can legitimately write about *anthropos* (self) in other than anthropomorphic terms. At any rate, Freud (1923) says:

> . . . We have formed the idea that in each individual there is a coherent organization of mental processes; and we call this his ego. It is to this ego that consciousness is attached; the ego controls the approach to motility—that is to the discharge of excitations into the external world; it is the mental agency which supervises all its own constituent processes . . . (S.E. Vol XIX, p. 17)

The image of Freud's realization of the ego as a prefrontal brain system structuring system-consciousness appears in a pictogram that shows the ego inhabiting an image of the brain. There can be no doubt that Freud is conscious of this anatomical localization of the structural components of the mind, for he tells us that

> the ego wears a 'cap of hearing'—on one side only, as we learn from cerebral anatomy . . . (S.E. Vol. XIX, p. 25)

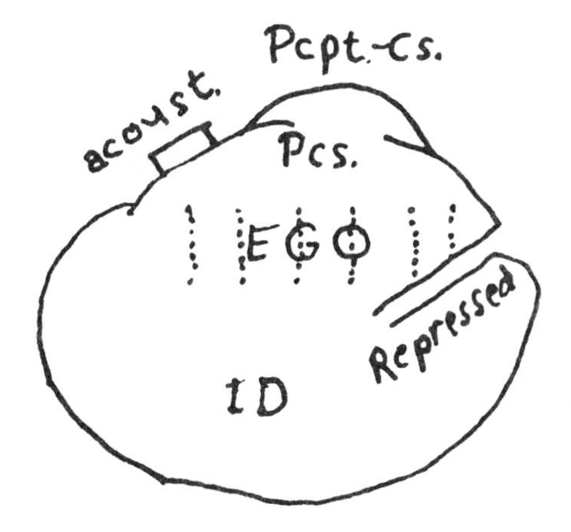

Figure 1

Despite the joking disclaimer "the form chosen has no pretensions to any special applicability" (p. 25), this diagram of the brain clearly indicates that the repressed appears below the sylvian fissure—where we would expect it as the repository of dynamically inhibited memories and fantasies that surround the amygdala. The ego itself occupies the anterior prefrontal areas of this brain image. Thus, Freud pictures the ego as anatomically rooted in the system perception-consciousness.

In *The Ego and the Id* (1923), the ego's perceptual functions determine its nature. These perceptual functions include both the perception of external objects and the so-called *endopsychic perceptions*, which here include pleasure and pain organized as feelings as well as word images. The perception of the body itself formed the core organization of the ego.

Freud's conception of a body ego was indeed firmly linked to a very physical conception of the brain.

> The ego is first and foremost a bodily ego; it is not merely a surface entity, but is itself the projection of a surface. If we wish to find an anatomical analogy for it we can best identify it with the cortical homunculus of the anatomists, which stands on its head in the cortex, sticks up its heels, faces backwards and, as we know, has its speech-area on the left hand side. (S.E. Vol. XIX, p. 26)

The brain's structure was thus in Freud's mind as he described the structure and function of the ego-self. Indeed, his anatomical placement of the ego showed foreknowledge, for, as an organ of internal perception, the tertiary cortex does owe its extended organization to the cortical homunculus. The premotor areas that have evolved into the prefrontal system for assessing motor sequences and grouping patterns have themselves evolved from the sensorimotor homunculus. We can understand Freud's idea of the ego being first and foremost a body-ego if we understand that each form of perception originates in a different distributed zone and system of feedback from the body.

To put the case in paradigm terms, the identity function of (1) agency perceives the body as sequenced motor images that it composes into an integrated tool for pragmatic action. (2) The subject function generates consummatory motives out of such perceptions as thirst, hunger, taste, smell, and sensory feedback from the oral, anal, and genital psychosexual zones. (3) The object function perceives and organizes postural images of the body through feedbacks that include (a) muscle and joint positions, (b) vestibular information, (c) somesthetic body surface information, and (d) auditory and visual data on space localization. The self function perceives the body's insides, which it composes of pleasure and pain signals, often relayed through the autonomic nervous system. "Feelings," the feedback from affect discharge, are essential components of this information.

Anchored in the body, the integrated identity functions operate simultaneously to synthesize a unified consciousness. What is distinctly human is the hierarchical development of the agent and object functions through sequenced speech and through visual fixation groups. Higher level ego functions must be developed to deal with this layering. I have been calling the highest level functions the identity functions. Freud included these in his concept of the ego. He used *judgement* to describe the biologically mandated synthesis of identity functions.

Writing about the development of Freud's concept of the ego, Heinz Hartmann (1964) shows that Freud's structural theory of the ego emphasized the biological roots of both the drives and the ego functions.

> . . .while in his earlier writings the drives had often been referred to before as "the biological aspect" of personality, now the powerful triad of functions: adaptation, control, and integration (synthetic function), attributed to the ego, underscored its significance as a biological agent . . . especially the conceptualization of the ego as, we may say, an organ of "centralized functional control" brought it closer to the thinking of brain physiology. (p. 290)

This reading of Freud parallels the present paradigm's holding the synthetic function as the coordinator of all the component identity and underlying ego functions, for Freud's attribution of hierarchical function to judgement comes close to the concept of a synthetic function. The paradigm might therefore call the human identity synthesis that unites products of the two hemispheres *judgement.*

Historically, Hartmann's own delineation of self and object representation as a set of coherent functions within Freud's ego opened the door to new theories of self. In this movement, Heinz Kohut may be taken as the speaker for what I call the *subjective-self* theorists (a group that includes John Gedo), while Otto Kernberg represents the theorists of the *objective-self.*

The major self theorists have chosen to amplify one or another aspect of identity as the central focus for the whole organization of mental process. Gedo and Goldberg (1973), for instance, emphasize an epigenesis of self organization, which they characterize as a hierarchy of personal aims. [This, I have been calling the agency function] (p. 10).

Kohut (1977) presents a more complete (but less organized) account of the subjective side of identity development, for his account includes both agent and subject functions. Kohut's understanding of *narcissism* focuses exclusively on the side of mental life that includes these functions. In truth, both aspects of his 'bipolar self' exist entirely in the dominant hemisphere.

> . . .With the term "tension arc," however, I am referring to the abiding flow of actual psychological activity that establishes itself between the two poles of the self, i.e., a person's basic pursuits toward which he is "driven" by his ambitions and "led" by his ideals. (p. 180)

Thus, his understanding of *narcissism* extends and amplifies Freud's concept of narcissistic libido into something the paradigm would call a complete theory of subjective development. While it emphasizes the creation of action plans, however, it neglects the integration of reality programs. The tension between Kohut's two poles is no more than the tension between ego tendencies and the tendency of the ego ideal, and it is no substitute for psychoanalytic drive theory.

Kernberg uses the term *self* to refer to the sum total of self and object representations. He builds these representations out of affectively connected self and object images into a concept of personality integration as character structure. His theory focuses almost exclusively on the identity functions of self and object. This means that his theory relates only to the nondominant hemisphere. While there is a good deal to be learned from these writers, there is a limit to the usefulness of theories that unconsciously neglect the fact that consciousness and its dysfunctions both exist inside a bicameral, lateralized brain that operates through the synthesis of its basic functions. What follows then is discussion of the impact of new brain information on psychoanalytic theory. These new information allow us to begin to fulfill Freud's dream of erecting a psychoanalytic theory on sound neurological information.

The paradigm's contribution to this deepening of psychoanalytic theory is to offer a solution to the ongoing problem of the relationship of drive to structure. Freud offered a first partial solution to this problem in 1914 when he postulated two kinds of cathexis, narcissistic and object cathexis. The notion of cathexis as a bonding of energy and structure was absorbed into his later ego theory so that in later theory the ego wielded cathexis. Hartmann's (1939) introduction of the notion of neutral and neutralizing energy broadened the scope of drive theory. Many psychoanalysts used this terminology to discuss the concept of libidinal and aggressive, narcissistic and object cathexes, as well as a neutralized form of each of these cathexes. The terminology became so abstracted as to defeat its usefulness. The paradigm returns the notion of cathexis to its origin in brain structure.

To translate the drive terminology into neuro-language, I identify Freud's ego with the synthesized operation of the identity functions. Drive is the effect of specific neurotransmitter systems on the major identity functions, and cathexis is therefore the actual dendritic contact between the drive neurons and the system-consciousness zonal neurons. In paradigm terms then, drive is the neurotransmitter that flows into neurostructure.

The brain-based drive theory lends itself to a different terminology than the one Freud used. While *libidinal drive* (as a life actuating growth energy) is a fair approximation of what I have called action drive, *aggressive drive* is a less useful name for the reality drive. Hartmann's neutral drive is fairly close to the meaning of neutral drive in the paradigm. The present terminology derives from an understanding of the identity functions of the medial zones. Since the medial identity zones are the tertiary assessment cortex for the limbic system, and since their functions are much more directly energy related, we must conclude that the major regulation of drive cathexis occurs within the medial zones. This is because the monitoring of consummation pressure and the assessment of the program of action plans are both medial functions that determine the amount of drive energy that will be made available for the lateral unfolding of action plans. Thus, the paradigm concept of drive cathexis leads us to some clinically useful conclusions. The action drive is monitored in the subject zone as subject-esteem, i.e., the degree of confidence that can be placed in consummation process. The action drive is transmitted to the agent zone as the degree of effort available to animate action plans. The reality drive is assessed in the self zone as self-esteem, i.e. the belief that frustration can be negotiated. The reality drive is assessed in the degree of concentration available for necessary feature detection in the search for salience.

EMOTION

What relationship connects drive, emotion, and anxiety? Drive has always been intuitively linked to emotion. Freud could not decide whether anxiety should be included within the conceptual context of drive or emotion. Clearly, anxiety and emotion both have a regulatory effect on system-consciousness. The question is, what kind of effect do they have?

Freud (1926) suggested that emotion and anxiety both have signal value. In *Inhibition, Symptom and Anxiety* he wrote that in the adult, anxiety functions as a signal to the ego, although as a signal that is not necessarily conscious. He could not decide, however, as appendices to that paper show, whether anxiety links with the ego's pleasure and pain regulation. Brenner (1982) treats affects (depression, for instance) as having signal value. No one, however, has distinguished precisely between the regulatory function of emotion and the regulatory function of anxiety. Like the concept of cathexis, the concept of signal value introduces too much ambiguity into analytic theory.

The solution lies in brain structure. Pleasure-pain (experienced as emotion) regulates system consciousness's medial zones. The anxiety systems regulate system-consciousness's lateral zones. The experience of such full scale anxiety discharges as rage or fear have feedback as emotional consequences. Thus, the emotional signals, especially shame and guilt, also operate to forestall the outbreak of anxiety.

We must consider a separate form of emotional regulation for each hemisphere. We have already distinguished *emotional expression* as a dominant hemisphere form of intrapsychic signal reception and interpersonal discharge from *affect response* which is a nondominant hemisphere form of intrapsychic signal reception and interpersonal discharge. The action hemisphere has an expressive nature. Gesture, whether hand or body, and facial expression (right-sided) indicate the state of the action system's readiness for consummation and continuation in action plan sequence. The nondominant hemisphere responds to reality. Its affects, conveyed in the left side of the face, or body, or hands, indicate the degree of frustration of action plans, or the degree of pain reality imposes. Because the right half of the brain controls the left half of the individual, the left side of the face manifests affect disposition and the degree and kind of difficulty in finding familiar salience in the environment. Of course, affect and expression are coordinated. For instance, the right hand may gesticulate, while the left hand shows limits and boundaries of behavior. Clearly, emotion—expression and affect—has both an *intra*psychic signal value that relates to the progress of action plans, and an *inter*psychic, interpersonal signal (or communication) value. This latter function is built up from instinctual predispositions for object relations to systematized forms of communication.

Expressive and affective signals indicate the progress of action plans. Action plans consist of a sequence of segments. For completion, each segment requires a particular indication of salience. This indication produces a relief of some of the drive pressure that maintains intrapsychic work. The relief must therefore be signalled along the pleasure axis, while unexpected increases in drive pressure must be experienced as painful. Presumably the completion of an action segment releases action energy that signals confident anticipation that further work will complete the entire program. Similarly, completion of salience finding for a segment of the action program must result in a liberation of reality energy that produces a reduction in frustration. In this scheme the completion of the whole action plan must yield a state approaching Freud's *nirvana* that includes all feelings of satisfaction and pleasure with contributions made to this state from each hemisphere. In general, the nature of the satisfaction must depend on the particular program that is achieving completion.

Orgasm is the prototype of one kind of emotional discharge. This is usually accompanied by a spindling response in the right amygdala. The fact that orgasm usually appears to be accompanied by right hemisphere deep limbic responses indicates that the release of drive pressure overall is controlled by the right hemisphere. If we include foreplay in the sequence that leads to orgasm, we can offer an explanation for a phenomenon that puzzled Freud. Freud wondered why foreplay did not produce pain rather than pleasure inasmuch as each segment of the foreplay increased the pressure for orgastic discharge. I think the answer is that the relief of each segment of the action sequence insures the greater imminence of orgastic discharge, and this produces a pleasure response triggered by each hemisphere—anticipation in the consummatory hemisphere, and relief in the affect hemisphere.

Anxiety discharge, even instinctualized anxiety discharge, is separate from the emotional system. Anxiety signals that do not reach consciousness correct the course of action plans during their unfolding. Conscious anxiety invokes conscious memory to correct or modify action, or to search for further salience. When, to take the dominant hemisphere as case in point, a sustained signal of error anxiety continues in spite of correct action, and when conscious anxiety increases, the result will be *fight-anxiety* and fight discharge—the rage response.

I believe that the narcissistic rage response is explained in this way. Anxiety discharges such as flight and fight or rage and fear (which are similar to the discharges found in animals) are instinctualized responses that disconnect the system consciousness (ego) so that they take place outside of its organizing context.

The emotions do not belong to the anxiety signal system. Emotions are instinctually object related. Anxiety discharges inhibit object relations to halt action plans. When the entire action machinery becomes involved in fight or rage response, object relations cease. It follows that it is misleading to group rage with frustration, irritation, and anger as coderivatives of an aggressive drive. Another way of saying this is that such emotions as love and hate are dominant hemisphere forms of emotional expression, while rage and fight are dominant hemisphere discharge anxieties. Of course, the emotional system and the anxiety system affect each other in the aftermath of their operations, and this produces a coordination of the two in the regulation of system-consciousness operations. Thus, a discharge of rage will have an aftermath of emotional expression that may include hatred and guilt.

SYSTEM-CONSCIOUSNESS OPERATIONS

The functions of system-consciousness are maintained, synthesized, organized, and regulated according to the basic structure of the system. In the adult, the highest level in the system-consciousness hierarchy is a conceptual and reflective structure of identity that also exists as the social side of the mind so that conceptual-reflective cortex and social cortex operate in a single conceptual framework. We experience this union through ideals and values.

Each stage of life organizes an identity format that conforms to the social-biological requirements of the era. New identity as it is constructed in the era forms the uppermost system-consciousness program. We may call such a program for overall synthesized identity an *identity-construct.* It follows that certain action programs are fundamental to the maintainance of the identity construct. These programmed action plans must not violate the essential construct of identity that maintains coherence in intrapsychic synthesis. Given these constraints, an individual develops a repertory of action programs suitable for the diverse needs of living.

The action program itself is composed of a dominant hemisphere portion and a nondominant hemisphere portion. Take the adult system-consciousness as an example. The dominant hemisphere portion is ordered as a proposition: 'I can do thus and so'. The nondominant hemisphere portion qualifies the action plan by requiring the presence of certain salient conditions. Action programs contain a series of propositions and conditions.

We have seen that anxiety controls the entrance of memory into system-consciousness thought. We have also seen that emotion demarcates and indicates the nature of progress in action-plans. Now we must add that in the adult system shame and guilt form the high level emotional signals that regulate the dominant and nondominant hemispheres respectively. Shame signals determine the degree of confidence that can be allocated to the overall continuation of an action program. The degree of shame indicates the degree to which the action program is necessary to the fulfillment of subjective identity. Ambitions and ideals are regulated by shame. Similarly, guilt signals regulate the self-esteem available for continuing processing and participation in social reality in the pursuit of action plan completion. Guilt determines the degree of force one may exert within reality and the social field to complete action plans.

Shame and guilt, the twin arbiters of the two hemispheres' highest identity functions, are the repositories within system-consciousness of the society's ideals and morals. Kohut's "tragic man" cannot fulfill his action plans because there is insufficient mirroring of his purpose in society for him to be able to maintain his goals. The "guilty man," made famous in the first half of the twentieth century, could not allow himself to continue his action plans because they involved damage to society. In different periods, changing social forces seem to emphasize controls in one or the other hemisphere, and re-enforce identity structures in one hemisphere more strongly than the other.

Dynamics and the Concept of the Unconscious

Our paradigm conforms to standard psychoanalytic understanding of the genesis of conflict in human life. Some analysis of the neurological meaning of terms does, however, make a number of classic definitions more precise. To begin, we must define fixation as a dominant hemisphere regulated process and trauma as a nondominant hemisphere regulated process. Fixation refers to a deeply engaging experience of a new mode of satisfaction or consummation. Presumably the kindling set off in the limbic system through unexpected gratification organizes a continuing set of experiences that gives rise to new tendency and important new action plans to repeat the experience and to make it part of the subjectively initiated repertory.

Similarly, trauma may be seen as an unexpected failure in gratification, or as an unexpected or persistent impingement from the side of reality that finally produces a kindling response and a deep engagement with painful experience. This deep engagement removes previous salience from the field of participatory expectations, or else it triggers an adaptive process that forms a new object-self or object-representation that reorganizes personality. As we know, during the process of trauma individuals will finally become removed or detached so that they appear to be involved with the object-self components of their own identity as if they were only social observers. This response to trauma is a stage in the adaptive reintegration of the personality.

Experiences of trauma and fixation are thus the organizers of long-term, generically useful memories. They organize new experiences and provoke the formation of hierarchically new identity constructs. As the life developmental process goes on, consecutive experiences of trauma and fixation overlay one another to produce the epigenetic psychic layering that gives rise to the dynamic unconscious.

Together with standard psychoanalytic theory, the paradigm holds that new periods of development are ushered in by inevitable psychic conflicts that are the result of intersecting biological and social forces. We have seen that human epigenesis causes inevitable trauma and fixation. The kindling that occurs during the period of formation of new psychic structure mandates the reception of new experience that forms new long term memory over a period of days or weeks. During this period of organizing experience, the volume of new experience triggers the formation of new action plans and the synthesis of new identity constructs. Themes of conflict between the old action plans, and the newly engendered ones produce a dynamic struggle that inhibits the old action plans and their identity regulators, thus forming a layer of dynamic inhibition. We call the action of such defensive systems primal repression, reaction formation, regression proper, or negation, according to the epigenetic level in which a new defensive system works.

The mental structures that channel the drives by containing the fixation and trauma through such operational defenses as condensation and displacement yield basal fantasies. The defenses make the new experience of the era tolerable and manageable. Operating in each universally experienced, biologically mandated epigenetic level, the defenses of each new period engender fantasies that are at once basal and universal. While the content of these fantasies differs for each individual, their form and basic organization must be almost identical. That is because there is, in each epigenetic period, a central organizing fantasy that channels the drives, both as to the consummatory aims and as to the manner in which one pursues these aims in reality. The basic organizing fantasy in each period of development is an elaboration of, and contains within it, earlier more nuclear forms. To give another example, adolescent masturbation fantasies further develop the organizing oedipal fantasies. The dominant hemisphere contributes consummation aims to this fantasy which are perceived as provisional, subjectively derived action plans. The nondominant hemisphere contributes the salients that would be necessary for the fulfillment of the imagined action plan. The so-called family romance fantasy often consists mainly of elaboration of salients—almost to the exclusion of action plans; which is to say that the person will present his or her fantasy in terms of a picture of the object-self as in possession of the attributes (rather than the actions) of the hero of romance.

New identity is built up epigenetically in biologically mandated *universal constructs.* These are formulations of identity that synthesize new goals amd new object-self representation for the new period of development. Because drives are medially regulated and channelled, universal fantasy issues from biologically mandated new medial brain/mind structure. Because pragmatic action and salience detection are laterally regulated, universal constructions of identity emerge out of biologically mandated new lateral brain/mind structure in the course of epigenesis. Both drives (which are medially organized) and their derivatives (which are lateralized) are subject to conflict, and this conflict takes place inside the framework of zonal organization. Our notion of conflict must include (1) the dynamic layering of new identity and new experience that sometimes conflicts with the old—*vertical conflict,* and (2) conflict between the two hemispheres—*horizontal conflict.* The notion of conflict between the two hemispheres takes into consideration the fact that new psychic structure is sometimes instigated by one hemisphere, and sometimes by the other. Sometimes there is only fixation, and sometimes only trauma at the beginning of psychic change. Thus, in the case of trauma, newly formed identifications may override the subjective goals that are still maintained in the dominant hemisphere in such a way that ambitions, ideals, and previous aims are no longer compatible with new objective identity structure. To take an extreme instance, consider the case of the heiress who was kidnapped by a radical militant group. This woman underwent forcible identification with these aggressors that changed her personality structure in a way that was incompatible with her old aims and ideals.

Horizontal conflict engendered in this way requires a change in subjective ideals and aims—what we may call brain-washing in this example—for its resolution.

To consider an example of dominant hemisphere induced change in identity structure provoking conflict, some periods of life usher in new forms of sexual gratification, and these require a revision of objective identity structure. What provokes guilt in one stage of life, does not provoke guilt in another. We can presume there is always some dynamic conflict inhering in the relation of the two hemispheres during a period of life, for life change, growth, and adaptation take a different pace in each hemisphere. Working out this conflict takes up a portion of mental life and mental energies. Error and novelty anxiety develop coordinated experiential signal value that differentiates in each stage of life development. Thus, we may see separation anxiety as the dominant hemisphere form, and stranger anxiety as the nondominant hemisphere form of signal anxiety in the six month infant stage of system-consciousness. Clearly the evolution of these functions of anxiety takes a different pace in each hemisphere affecting the nature and timing of what becomes conflictual.

The organization of new experience around pleasure and pain processes during kindling forms dynamically charged unconscious nuclei of experience. These nuclei are close to the neurological sources of instinctualization of the system-consciousness, and they therefore operate outside of the framework of system-consciousness, for system-consciousness provides a frame that protects against instinctualized behaviors. When system-consciousness expands to include new kinds of experience within its frame, the new experience must be habituated and conditioned, made capable of being framed in thought and no longer capable of evoking the kindling process.

The system-consciousness operates to insure the successful completion of action plans. Any process that causes the disruption of the smooth flow of thinking must be kept out of consciousness. We may think of repression as the model of defense then, as a set of inhibitory systems that are set into motion by signal anxiety to maintain the barrier against the stimulation of subcortical limbic sites that interrupt the smooth flow of system-consciousness. Error and novelty anxiety form in layers. We must presume that the system of defense is organized in layers that correspond to the layered development of system-consciousness.

In addition to the dynamic unconscious, a second type of unconscious corresponds to the *selective-attention-function* of the ego. Freud referred to this function as "hypercathexis." Presumably this function operates with neutral drive energy. Thus neutral drive energy is utilized in the selection process as system-consciousness changes its distributed links during the course of its operation. We may think of selective attention as determining the pattern of information that are excluded and included in potential consciousness in each identity zone. *Denial-of-reality*, for instance, may be seen as a lateral nondominant hemisphere operation. As an ordinary selection mechanism, denial excludes what is not salient. To give an example, in walking down the street we see only those landmarks that have salience to us because they signify possible importance in our action plans.

Similarly, the lateral dominant hemisphere has a process of selection for determining what shall reach verbal consciousness. Freud and Luria say that voluntary processes are made conscious through verbalization, which means that we become aware of our action plans through verbalizing them. Verbalization is the first step in turning potential action plans into actual plans. Verbal selection—hypercathexis—is an aspect of the dominant hemisphere choice of what will attain consciousness.

There appears to be little doubt that emotions are experienced directly. So too are pain and pleasure, the inner chemical introceptive modalities of smell and taste, and body discomfort and organ sensations. It appears that these are medially processed forms of sensation that reach consciousness in an off-on mode. One either feels emotion, or one does not. The most profound grief may be completely suspended for a moment before it resumes. The countercathexis against intense emotion must therefore employ high levels of energy. Mild emotions, however, may be sampled using hypercathexis of the emotion, and verbal equivalence of these forms of emotional consciousness may be used to explore the emotional state, for the system-consciousness uses a verbal system to synthesize consciousness gathered from all four zones.

The Process of Change

Psychoanalytic theory has developed a set of terms to describe the stages and substages involved in the basic process of mind-brain change. Language applicable to the regressive phase comes from Ernst Kris, who coined "regression in-the-service-of-the-ego." In *The Interpretation of Dreams* (1900), Freud himself introduced a vocabulary to describe both the regressive and progressive phases. I have used Kris's and Freud's terms to describe the basic process for structural change in system-consciousness. This process is the same as that in (1) the epigenetic growth and adaptive processes, (2) the creative process, (3) the psychoanalytic process, (4) the dream process, and (5) syndrome-formation process. The stages follow the same order, moreover, whether they occur over a prolonged period of time as substages in a larger movement, or daily, as they do in dream formation.

The dream process as Freud described it is the prototype for this process of change. We may also look upon dream formation as accomplishing the nightly revision of action programs. In the first stage, which takes place during the day, a highly cathected action-intention cannot be carried out because of interference. The failure to carry out the intention arises through stimulation of a conflict. The conflict mobilizes anxiety and a loss of the hypercathexis of the verbal intention. The verbalized intention loses its signal of inclusion in distributed system-consciousness, yet it is not abandoned, for it continues to maintain a relationship to the medial formation of wishful drive cathexis. Thus the verbal intention maintains its action pressure. At night, after REM sleep has begun, the selection pressure against the verbal element drops away. Now, the anxiety system need not operate because the voluntary action system is disconnected in dreaming sleep. The verbal derivative of intention is therefore free to move back through the layers of mental organization to earlier forms of verbal and gestural representation. This regressive phase of the process goes on while the wish for consummation is still active. Finding representation in the nondominant hemisphere, the verbal residue as it has been transformed through regression to more primitive modes of rendering flares into visual imagery. This phase must take place in the context of memory consolidation, for the visual imagery that transforms the latent verbal content uses day residue visual imagery that was denied at the moment when the verbal initiative was dropped. In other words, the verbal initiative was dropped from system-consciousness at the same moment that anxiety forced selective depletion of salient features from reality-consciousness. The transformation of the regressed latent content into imagery of an earlier period makes use of the day residue of denied salient features.

The disconnection of the hemispheres during the dreaming process, and the disconnection of voluntary motor activity and of the sensory processing system during the REM dream state allows conflictual material to be processed in memory without setting up those instinctualized responses that could lead to kindling and untoward change or behavioral loss of control if the particular experience was allowed to consolidate during the day. Thus we can imagine that at the center of the dream transformation process kindling—amygdalar discharge—may occur, harmlessly, with only the formation of dream imagery rather than the forced reception of new experience, which would happen during the day.

At this point the dream imagery begins to take on significance in the nondominant hemisphere sphere of representation. The progressive phase of the dream process moves up through layers of mental organization, giving the dream imagery renewed causal connection. This progressive portion of the process may be equated with transference, for as Freud originally defined it transference is an intrapsychic mechanism for projecting earlier modes of mental organization on to the present. This progressive transference finally embodies a complete representation to the dream, which now exists as a mental product in its own right. As the dream must relate to the continuing action plan concerns as it emerges into wakeful consciousness, its manifest content is related to the action concerns of the following day.

In summary, then, the process for change consists of (1) a phase of problem-solving failure with attendant conflict and anxiety, (2) a phase of regression, (3) a phase of transformation, (4) a phase of progressive transference in which the imagery is revised, and (5) a phase of secondary revision in which the dream becomes a mental product equivalent to a new product of consciousness.

Epigenesis

I: The Developmental Framework: Infancy (0–18 months)

Because the basic organization and process for consciousness is a highly stable entity, the brain/mind apparatus remains stable throughout life. Hierarchical layers of superordinate control develop as epistructure that modifies, but does not alter, the basic functioning of the mental apparatus. This basic apparatus consists of (1) a thalamic, integrated and synthesized immediate memory mechanism and (2) a forebrain evaluation mechanism. In their conjunction resides the true organization of consciousness.

Thalamic regions are embryologically linked to corresponding higher cortical regions through the migration of a common type of columnar cell. The thalamic areas receive information for preliminary integration from both the limbic system and the sensorimotor system. The corresponding higher cortical areas assess this information and select further neuropsychological functions that determine which avenues of information will be pursued further, or else they determine a discharge process on the basis of the assessed information. Immediate memory (the content of system-consciousness) consists of those bits of thalamic information that can be held in conjoint synthesis.

The stability of system-consciousness depends on the relationship between the momentary information held in consciousness and the apparatus for evaluating, assessing, judging, selecting, and deciding on it. The feedback operations between the thalamic and prefrontal levels are increasingly dominated by conscious psychological processes as life development and epigenesis proceeds.

The relationships are as follows. The discharge decisions of the apparatus are based on limbic signals that are assessed in the medial prefrontal areas. The neuropsychological selection functions are based on the evalution of sensorimotor information carried out in the lateral prefrontal areas. Thus the thalamus presents dominant hemisphere medial consummation pressures integrated with action sequence information. The nondominant hemisphere integration of information consists of an affect pressure integrated with reality information. This integration indicates that conditions exist in reality that will allow discharge to proceed to consummation. As Freud often said, *judgment* is the overall function of system-consciousness. Judgment assesses the equivalence between the two information sets. Equivalence trips the discharge process.

The capacity of immediate memory is relatively fixed. Early in infancy the synthesis of immediate information may last up to 15 seconds. Later it may last for perhaps 30 seconds. Then a new train of immediate information appears and is processed. At first, therefore, discharge consists of instinctually mediated release in the face of a perceived congruence between information present in both hemispheres. A variety of such innate programs insures the organism's biological viability. Gradually, throughout infancy, the programs develop some flexibility in which delay interposes between need releaser and consummation. Mechanisms for both emotional memory (conditioning) and cognitive memory (habituation) are introduced into the operational format for system-consciousness. For this reason, each stage of life has a different range of discharges and judgments; thus each stage of life develops its own identity level programs for the total synthesis of consciousness.

Throughout the first eighteen months of infancy, the basic apparatus matures to the point of its first major epigenetic revolution. In the first weeks of postnatal life the maturation of the cerebral organization of the autonomic nervous system produces states of time-limited facilitation of the system-consciousness. These autonomic states command system-consciousness in an identifiable sequence. The sequence consists of (1) an affective-reality state, (2) an attention state, and (3) a consummation-action state. For reasons already examined in this book, we may presume that norepinephrine fuels the reality state, that serotonin maintains the attention state, and that dopamine powers the action state. All three major neurotransmitters are thus employed in the maintenance of system-consciousness activity. The immediate increment or decrement of available neurotransmitter must be the determining factor in acceleration or deceleration of state-related processes of system-consciousness.

Infant States The *orientation reflex* commands attention. Brazelton (1980) describes an innate patterning of states that determines the infant's response to the world. According to Brazelton, in the first months of life an orientation reflex brings the infant to attention in the presence of the mother. Immediate mutuality of gaze facilitates the attentional state through an innate neural mechanism. A phase of accelerated motor activity follows the attentional state. This leads to a discharge phenomenon, or to a deceleration toward sleep or a repetition of the cycle. This state sequence typifies the autonomic nervous system's way of generating the sequential states that make up system-consciousness's basic operating mode.

The infant is thus driven by the instincts to respond to reality. Once the engagement is made, forebrain mechanisms insure interpersonal brain communication through the eye contact during the attentional phase. Eye contact with the mother insures the communication of needs. Then, motoric expression accelerates, pressured by the action drive to find the right combination of movements to engender satisfaction. Satisfaction follows maternal reciprocation—or else attention breaks off. This is the basic sequence through which immediate memory, attention, and extended consciousness are engaged. Each small episode—and Brazelton says there may be up to four a minute—is repeated through life. Each 15 to 30 seconds of our lives we attempt to solve some little or big, segmental or whole problem of gratification. We awake to reality until our attention engages. Next, we try potential action to solve the problem. Finally, we evaluate the consequences. Our momentary destiny is thus the product of the major neurotransmitter systems. We hook into reality with our norepinephrine system, then we neutralize with our serotonin system as we enter into thought, then our dopamine system builds up pressure, then this pressure too is neutralized by attention, and finally we must make one of the 1,000 to 2,000 decisions we make every day.

Brazelton's (1979) assessment of the infant is based on an assessment of each of the components that makes the basic sequence operational. Brazelton's observers are trained to assess autonomic discharge phenomena, orientation, response to stimuli, motor tone and activity, and states of consciousness relating to attention. The following passage indicates the nature of the observations of infant behavior they are trained to make:

> The newborn . . . can habituate . . . startle responses to the same bright light . . . He shows a set of visual and motor responses to a moving object which include alerting of his face and body, widening of the pupils and eyelids, opening of the mouth, softening and raising of the cheeks, raising of the eyebrows, inhibition of generalized body movements and coordinated tracking movements of the eyes with coordinated head turn to follow it . . . he can make preferential "decisions" about stimuli which can be quantified by measurement of the duration of attention and by a description of facial expression. (p. 185)

From this we find that the prototypes for system-consciousness include (1) motor readiness and tone (consummatory), (2) perceptual searching (orientation), (3) discharge regulation (judgmental precursor of reflection), and (4) expressive-affective response (experience).

There are autonomic physiological correlates of each stage of the state sequence. Accelerated heart rate accompanies the initial orientation. The cardiac rate decelerates when attention engages. As the state system matures, novel stimuli may elicit only a very brief cardiac acceleration before attention cycles in (Berg & Berg, *Handbook of Infant Development,* 1979, p. 318). As the specific sensory systems begin to mature in the second month of life, specific stimuli become interesting and invite precognitive information collection of stimuli. It is now possible for the infant to use more precise data in seeking to match need with gratification. Freud called this accession the development of practical thought.

In the first two months of postnatal life sensory systems are mainly the so-called secondary ones that perform in neurologically diffuse and global ways, collecting and integrating data in the midbrain as part of more ancient evolutionary circuits. The specific form of sensory data that supersedes these systems comes on stream in the second month of life. The smiling response to the human facial configuration that materializes in this period is an instinctual release mechanism that joins a cognitive instinctual nucleus with an emotional response. Early cognition and specific perception is based on instinctual predilection for the reception of particular configurations of stimuli. Berg and Berg (1979) cite evidence that the curvilinear lines of the human face automatically trigger the cardiac deceleration of attention in the infant over two months of age. The cognitive recognition process comes to be part of the coming to attention mechanism that inhibits the more global sensory and motor systems of the first two months of infancy. Similarly, in the auditory modality during this period a pulsed auditory tone in the range of the human voice is most effective in eliciting the attention response in young infants as early as the second month of life. Berg and Berg summarize this information:

> . . .the data indicate a potent and surprisingly consistent effect of pulsing auditory stimuli. With few exceptions, deceleration in response to auditory stimuli in infants six weeks of age and less have been reported only when stimuli were intermittent. . . (p. 285)

The symbiotic stage coalesces as the first complete system-consciousness when the specific modalities have been linked to the infant's emotional response to the mother. The symbiotic system is, indeed, primarily a means of interlocking the maternal and infant brain organization. The effect of mutual gaze is primary in this, for the neutralizing state of maintained attention that accompanies mental activity of a synthetic sort is increasingly guided by mutuality of gaze.

Osofsky and Connors (1979) discuss this point as gaze develops as a regulator of autonomic states and of cognitive and emotional processing in the first few months of life.

> In the visual relationship between mother and baby, the infant guides maternal behavior. Mothers tend to look where their babies look . . . In general, the infant initiates and terminates the mutual gazing. If the infant initiates a gaze, the mother will usually respond and return the gaze immediately. If the mother initiates a gaze, she will usually maintain it until the infant looks . . . The infant's intrinsic biological process of gaze alteration allows him or her to modulate and thus maintain an optimal level of stimulation . . . (p. 534)

Symbiosis (3–6 months) The structural paradigm looks at symbiosis as system-consciousness's first thorough integration and synthesis. This systematization of identity functions facilitates the coalescence of ego functions (such as body image formation) and it produces a complete mental (intrapsychic) system, in which the infant's sense of identity includes, and is included by, the mother's identity structure. Mother and infant divide the identity functions, between them: (1) the infant's dominant hemisphere and the mother's nondominant hemisphere and (2) the infant's nondominant hemisphere and the mother's dominant hemisphere form a single interactive, reciprocal brain system. The mother's reality response reciprocates the infant's action integration. The mother's action integration is reciprocated by the infant's reality response. The reciprocation is entirely necessary for the infant does not have the capacity to synthesize a complete bihemispheric set of identity functions without the mother's symbiotic participation.

We may infer that most of the infant's consciousness of identity resides in the action organizing hemisphere. I have called this form of consciousness "subjective" not because it is internal as opposed to external, but because it refers to an independent center of initiative. The subjective identity functions initiate pragmatic action. This infant's "gestural" consciousness corresponds with Piaget's characterization of the action-centered infant.

Piaget's and Mahler's formal designations of the stages of infancy are compatible, providing a general context for the changes in system-consciousness that occur during the substages of infancy. The first eighteen months of life constitute a sensorimotor period, according to Piaget, in which the infant fails to distinguish the reality of objects from his subjective action effort directed toward them. Winnicott (1965) sees this whole period of life as characterized by the "transitional illusion" that the infant's activity creates the object's existence. Winnicott argues that the experience of a "true self" occurs when the infant experiences his mother reciprocating his spontaneous gestures so as to satisfy his needs (p. 140). Together, the motor sequence and the pressure for consummation engender feelings of what I call "subjective esteem" as the mother fulfills her major role in the symbiosis. Winnicott's "false self," on the other hand, may be understood as the result of the mother's insistence that her gestures must be reciprocated by the infant. We may think of this as the other strand in the symbiosis. When the infant is able to accommodate the mother he is treated as the "good object," a state of objective being that the infant tries to reproduce. In this state the infant may feel what I call "self-esteem."

Thus, the infant's subjective action hemisphere links up with the mother's reality brain, and the mother's action brain links up with the infant's self and object nuclei. Sroule (1979) discusses how the smile adds to muscle tone as a signal from the infant to the mother that aids in monitoring the interaction of the infant's with the mother's brain.

> *Thus, the social function of the smile complements the function of positively toned tension release by providing opportunities for the infant to exercise its tendency to perpetuate novel stimulus situations.* The tension-release mechanism enables the infant to remain oriented toward novel or incongruous stimulation and maintain organized behavior. At the same time, the smile signals well-being and encourages the caretaker to continue or repeat interesting events. In addition, as a behavior each partner can exhibit, as well as elicit from the other, it has an important place in the learning of mutual effectance. (p. 500)

The effectance of symbiosis through the use of instinctual mechanisms is also limited and broken off through the breaking off of instinctual mechanisms. Thus gaze breaking has an immediate effect on the autonomic stability that accompanies symbiotic contact:

> Similarly, gaze aversion, as an expression of negative affect, is associated with a reduction of arousal in face-to-face play . . . Such mechanisms have also been suggested to be *the counterpart of the* behavior-arresting *orienting reaction* to novel events, releasing behavior following processing of the event. (p. 500)

Mutuality of gaze therefore maintains attention and the mutuality of consciousness. Gaze is a neurological function that is connected to a large number of cortical and subcortical areas. The mutuality of gaze in eye contact truly maintains interpersonal brain communication of a sort that simultaneously effects the autonomic nervous system of each partner. Clearly the symbiotic stage remains the basis of the nonverbal signal system throughout life.

Each hemisphere has instinctual prototypes of communication that function automatically to elicit response in the symbiotic partner. Thus far we have considered gaze and smile, both mechanisms for the maintenance and breaking off of mutual contact. Now let us consider the signals that function within each hemisphere to monitor discharge processes. Crying and laughter are the nondominant hemisphere and dominant hemisphere mechanisms of signal communication respectively that function to indicate the degree of drive tension in each system. What is more, crying and laughter develop into extended cognitive mechanisms for communication.

Crying and laughter are instinctual roots of extended communication. The rhythm, high pitch, and variable intensity of crying manifest qualities that inhere in affect display as communication. The prosodic affective melody within speech is conveyed in the vowel sounds of speech. The open higher pitch of vowels as compared with consonants is closer to the sound of crying. Luria (1980) asserts that the nondominant hemisphere processes the affective reverberation of language (pp. 379–380). Crying manifests the basic desiderata of the auditory modality to which the nondominant hemisphere responds. Vowel-like and affective, crying indicates frustration with reality, for crying is a physiological obliteration of reality and a denial of reality signals.

Laughter is the prototype of meaningful expressive speech, containing the break in tension and the discharge which is the goal of every action plan. Thus, like crying, laughter manifests the drive system of its own hemisphere. The instinctual roots of laughter in the motor initiation system can be seen manifested in the catalepsy syndrome where a person falls to the ground in paralysis after a laughter trigger. This phenomenon is often seen in normal people when they become weak with laughter, giving rise to the expression "to die laughing." The action drive discharged in laughter gives rise to the semantic roots of speech. The phoneme characteristics of speech, a burst of separated consonants, find prototypical expression in laughter. The "ha-ha-ha" of laughter shows the consonants bursting in sequence, as opposed to the drawn out higher pitched melody of crying.

When the tension of expectation with the caretaker reaches a certain point, or when the striving for satisfaction is acknowledged by some communication with the caretaker, then as early as four months laughter can ensue as expressive-motor discharge. As the building up to laughter goes on, there is physiological indication of increased attention (cardiac deceleration) until at the point of laughter cardiac acceleration begins, and gaze and gesture contact are broken (Sroule, 1979, p. 505)

Sroule (1979) writes:

> . . . a concept of tension fluctuation is required to account for positive affect. When young infants break from a captivating event, they not only avoid distress with the modulation of arousal, but also frequently express positive affect. Also with many types of events, laughter occurs at a focused termination point. (p. 506)

This echoes Freud's (1905) conclusion in *Jokes and Their Relation to the Unconscious* (S.E. Vol. VIII) that libidinal energy is released when the unconscious libidinal tension of the wish is released. In line with the terminological distinctions I have made to distinguish separate hemispheric regulation, I would refer to laughter as an expressive rather than an affective discharge.

Crying and laughter are the prototypes of the emotional medial system for discharge regulation. We may consider the expressive emotions as dominant hemisphere forms of discharge regulation signals, and the affective emotions as nondominant hemisphere forms of discharge regulation. In this light we can distinguish between rage and anger, where rage is a reaction to blockage of the expressive action system and anger is a response to frustration imposed by a failure of reality to provide the proper gratifiers. Sroule (1979) discusses rage as a prototypical emotion:

> Shortly after the newborn period, head or limb restraint is apparently an adequate stimulus for extreme distress . . . only when the *rage* is due to "disappointment"—the failure of a *motor* expectation or interruption of specific ongoing activities—is it considered an emotional reaction in the present scheme. The reaction derives from the infant's involvement with a *particular* event. For example, the failure of a well established action sequence to be continued . . . (p. 487)

Development of Emotional System in Infancy Although many prototypical instinctual, even communicative, emotions arise in the first months of life, the emotional system becomes functional as a signal system only with the advent of symbiotic consciousness.

The ability of attention—system consciousness—to modulate and regulate the drives, subject them to modifications, and set them to work as instigators of specific ego functions that are coordinated aspects of the action program is known in psychoanalytic terminology as *neutralization,* here called neutralization of the action drive and neutralization of the reality drive. At the beginning of the symbiotic period a short term memory mechanism utilizing signal anxiety begins to become a part of the system for stabilization of attention. Prior to this period anxiety merely produces a break in attention without a means of signaling a restabilization.

Like the emotional system, the anxiety system is hemispherically organized. At around five months, separation-anxiety begins to determine whether action-plans and symbiotic consciousness may be pursued. At around the same time, or a little later, stranger or novelty anxiety begins to become a signal force in determining whether the conditions exist in reality for a continuation of symbiotic consciousness and brain exchange. The two forms of anxiety, action anxiety (separation anxiety) and reality anxiety (stranger-anxiety) become a safeguarding system for the maintainence of cognitive functioning. As signals they effect the lateral selection system of ego and neurophysiological functions in such a way that if either form of anxiety is signalled, then memorial processes specific to the hemisphere are invoked.

The overall functioning of system-consciousness as an intrapsychic system of unified identity synthesizing emotion, cognition, memory, and anxiety during the symbiotic period is evident in the following quotation from Emde and Gaensbauer (1976):

> We thought that comparing faces reflected some kind of internal cognitive matching process. During the first month or two of its appearance, the infant's facial expression accompanying this behavior seemed to communicate interest; it seemed like a manifestation of problem solving rather than pleasure. Later a fascinated look seemed colored with an uncertain negative quality, which often became a "sober expression." To our surprise, film ratings of "sobering" illustrated a definite developmental onset before the intense frowning, fussing, and crying of stranger distress. The onset of sobering was concentrated in the five to seven month period. (pp. 97–98)

Novel perceptions either increase attentiveness, i.e., produce *concentration,* or result in affect discharges that have either a signal significance or are so strong as to break off connection with the novelty through the behavioral state change induced by the autonomic discharge.

As the symbiotic consolidation period begins to come to an end at six to seven months a longer term form of memorial process is invoked increasingly to maintain the stability of system consciousness. During the so-called practicing phase when motor and perceptual skills increase dramatically as they are stabilized by memory re-enforcement, symbiotic consciousness tends to become more of an intrapsychic system and less of an interactional system. The place of the reality mother in consciousness is increasingly taken by a mental product—the image—or transitional illusion. The infant is able to invent new action schema that result in reappearances of a variety of familiar reality objects that are capable of substantiating the transitional illusion (the infant's omnipotent belief that initiation produces the manifestation of reality).

Mahler, Pine, and Bergman (1975) describe the infant's exploration of novelty and familiarity as symbiosis with the mother begins to give way:

> At about six months, tentative experimentation at separation-individuation begins. This can be observed . . . as pulling at the mother's hair, ears, or nose, putting food into the mother's mouth, and straining his body away from mother in order to have a better look at her. . .
> . . .There are definite signs that the baby begins to differentiate his own from the mother's body. Six to seven months is the peak of manual, tactile, and visual explorations of the mother's face, as well as of the covered (clad) and unclad parts of the mother's body . . . These explorative patterns later develop into the cognitive function of checking the unfamiliar against the already familiar. (p. 54)

Just as each stage of life, each formation of a layer of operational consciousness is mandated neurologically, so also is the breakup of that formation. The substages of infancy that Mahler and Piaget describe manifest first the coalescence of the symbiotic stage of the synthesis of consciousness, and then the substages in the disintegration of that coalescence. Thus, as separation-individuation sows the seeds of inevitable disruption of symbiosis, stranger anxiety and separation anxiety are invoked more and more often without producing a satisfactory maintenance of the symbiotic form of consciousness.

As we know from Piaget, the one year old develops new means and new goals (new action schema) with which to circumvent the constraints of action obstacles and action anxiety. In the face of separation anxiety the infant will respond with rage to the mother's or some other adult's failure to reciprocate the new action schema. This period of "practicing," often conducted outside of anxiety constraints through the mediation of a prototypical imagination based on treating the world as a sensorimotor or transitional illusion, finally gives way to extreme forms of anxiety with a dissolution of symbiotic consciousness as the organizer of the infant's world. Ainsworth et al. (1978) have described how two behavioral systems, one directed toward union with the mother, and the other directed toward exploration, eventually produce so much divergence that a new system of consciousness must come into being (p. 4).

As the system of symbiotic consciousness begins to give way, new islands of consciousness, i.e., new forms of perception and new action schema appear, separation anxiety progresses to rage and finally to fighting, and a kindling of dominant hemisphere action instincts. Similarly, as stranger anxiety gives way, anger and then flight operations develop, and these are accompanied by prolonged orientation and nondominant kindling. Thus, if the symbiotic period may be termed the period of first formation of a system-consciousness, then the rapprochement may be termed the period in which system-consciousness is first transformed.

We must be attentive to the process through which consciousness is transformed and through which a new layer of system-consciousness develops, for this transformation becomes a prototype for all further major developmental or pathological or instrumental changes in system-consciousness. In the symbiotic period all of the neurological roots of consciousness are tied into one mental system which is synthetic and coherent. In the rapprochement period of breaking down and then reforming a new system-consciousness a new element is added—the layering of consciousness. This layering provides psychological depth and a vertical form of defense in which a system of operations must be developed through which the old system and the new system maintain their boundaries and form a single overall system of mental and emotional life.

Mahler and her group describe the darting away and shadowing behavior that accompanies the transformation called rapprochement that takes place in the eighteen to twenty-four month period. System-consciousness aim-fulfilling mirroring of mother and of other children is broken by sudden darting away to separateness. The same rapid alternation manifests in peekaboo games, ended with intense temper tantrums. The newly sudden relinquishments of consciousness in going to sleep also attest to the ambitendency of the drives: two behavioral systems which are no longer capable of being welded in the same patterns of consciousness which served the six-month period and beyond. In discussing the rapid mood swings and the insatiable ambitendency of this rapprochement crisis, Mahler (1975) mentions an unexpected observation—a new phenomenon discovered by her group.

> . . .An unexpected and strange phenomenon appeared, seemingly a forerunner of the projection of one's negative feelings:

this was the child's sudden anxiety that mother had left, on occasions when she had not even risen from her chair! There occurred, more or less frequently, moments of a strange, seeming "non-recognition" of mother, after a brief absence on her part. (p. 95)

The Shattering of Infancy: The Primal Scene The trauma of the *primal scene* determines the shattering of infancy. As I use the term here, the primal scene is a construct. The term refers to a traumatic organizing event in experience that determines the beginning of the end of infancy. It is an inevitable event that shatters symbiotic consciousness, leading to a new formation of medial consciousness, and to a new stage of life in which system-consciousness is reorganized, ultimately to conform to the structure of language.

At around eighteen months the infant faces an apparently insoluble challenge to its identity structure. Resolution involves the shattering of the transitional illusion that the infant is the maker and shaper of its own reality. The sensorimotor illusion that active effort creates the object shatters. A new form of objectified consciousness follows. In this consciousness the infant itself is but one individual person, like its mother and father and others.

In his conclusions to *An Infantile Neurosis* (1918) Freud makes these very points:

> . . . If one considers the behavior of the four-year-old child toward the reactivated primal scene, or even if one thinks of the far simpler reactions of the one and one-half year old child when the scene was actually experienced, it is hard to dismiss the view that some sort of hardly definable knowledge, something, as it were, preparatory to an understanding, was at work in the child at the time. We can form no conception of what this may have consisted in; we have nothing at our disposal but the single analogy—and it is an excellent one—of the far-reaching *instinctive* knowledge of animals.
>
> If human beings too possessed an instinctive endowment such as this, it would not be surprising that it should be very particularly concerned with the processes of sexual life, even though it could not be by any means confined to them. This instinctive factor would then be the nucleus of the unconscious, a primitive kind of mental activity, which would later be dethroned and overlaid by human reason, when that faculty came to be acquired, but which in some people, perhaps in every one, would retain the power of drawing down to it the higher mental processes. Repression would be the return to this instinctive stage, and man would

> be paying for his great new acquisition with his liability to neurosis, and would be bearing witness by the possibility of the neurosis to the existence of those earlier, instinct-like, preliminary stages. The significance of the traumas of early childhood would lie in their contributing material to this unconscious which would save it from being worn away by the subsequent course of development. (S.E. Vol. XVII p. 120)

Following Freud, I contend that the eighteen month old infant is instinctually prepared to encounter an organizing trauma. In this sense the primal scene is a *universal organizing experience.* The scene can be made up of any mutually exclusive engagement between the parents, when they are impervious to the supercharged action wishes and salient expectations of the infant. This readiness for trauma makes a necessary and sufficient organizing experience into a focal event that triggers massive hippocampal facilitation of new experience. Whether the infant witnesses a sexual event between the parents, or some other engagement, the effect is still the same: the shattering of the infantile illusion of sensorimotor omnipotence.

With Freud I assert that the original primal scene acts as an organizer of the reality sense. It initiates the identity function of reality consciousness. As the experience of the primal scene shatters the false illusion of reality that the symbiotic consciousness maintains, it also establishes a highly facilitated perception of reality as a promoter of psychic pain. From what we know about kindling we may conclude that such a complete trauma induces the intense forced perceptual intake that creates a complete organizer for long term memory formation of the generic or emotional kind. The event, itself buried, nevertheless stands out as the singular beginning of the object-self as spectator. Harris and Harris (1984) have reconstructed the primal scene construct centering Freud's identity development.

The emphasis on a new form of reality consciousness, on a figure that has emerged from a background, and that can be contemplated in consciousness as a representation, is the sine qua non of the new consciousness that is organized in the wake of the shattering of infancy. The experience itself is the prototypical neurologically mandated trauma, acting as an organizer for reality. Just as other traumas later in life can draw experience to their own organizing images and representations, so the primal scene acts as an undergound (amygdalar-hippocampal) channel drawing perceptual data and events into its own repressed field of organization, preventing, as Freud said, the wearing away of the primary organizing event by subsequent experience.

II: The Layering of Consciousness: Later Infancy (18–24 months)

Overview The rapprochement crisis heralds the end of infancy. In this period the formation of a newly synthesized system-consciousness inhibits the symbiotic consciousness of infancy that interlocks the infant's brain with the mother's brain. While symbiosis continues to exist as a mode of thought that synthesizes gesture and reality images, that is, as a synthetic ego function, it is generally subordinated at the identity level to representational consciousness. These developments entail emotional and cognitive changes, and they create the first major transformation of consciousness.

The new representational synthesis of intrapsychic consciousness organizes itself around two allied newly differentiated forms of object consciousness. These newly differentiated identity functions are the object-self which constructs the body as an object in spatial reality, and the object which I define as another person's representation available as a model of imitation. These newly differentiated object functions mediate the identification and introjection mechanisms through which object-self representations and object representations develop. Subjective identity functions also develop along representational lines. They are guided by a grandiose sense of possibilities of action programs that are sensed as capabilities of the parents. The parents are experienced as ideals of action planning. Thus, the agent function of identity organization expands along ideals of parental imagos of action.

The effect of overlaying a symbiotic integration of consciousness with a representational synthesis of consciousness produces a new mental quality of *inner-observation.* It is a momentary reflective sense of coordinated inner observation. The synthesis of a whole intrapsychic identity adds a sense of depth to each separate identity integer, subject agent, self, or object-self. Thus, the representational self conclusion "I feel," or subject conclusion "I want" comes to consciousness. Clearly the use of synthesized speech gives rise to the experience of an experiential level beneath the speech or thought giving rise to it.

The appearance of the sentence marks the transformation of consciousness: the sentence is the synthetic form of consciousness arising during early childhood. It integrates each hemisphere's information and synthesizes a whole product of consciousness—a representation in words. Let us consider some of the neurological and neuropsychological developments that make the representational form of language use possible.

Neuropsychological Developments Piaget's book *The Construction of Reality in the Child* (1954) informs us about the change in object perception and action schema before and after the rapprochement period. Piaget envisions the change from the fifth stage of sensorimotor development from twelve to eighteen months to the final stage of insight and object constancy as revolutionary. According to Piaget, before eighteen months the child is engaged in perceiving a field of reality which consists of a spatial container composed of all of reality. After this period, objects emerge in their own distinctive clarity as figures out of a background. According to Piaget, this development corresponds to the child's ability to maintain a mental representation of the object despite the disappearance of the object.

The concept that objects composed of specific features differentiate as distinct entities out of a whole spatial container relates to the development of the nondominant hemisphere as a representational information processor. Figure background distinction implies a mental capacity to perceive objects as content within a reality frame. The dominant hemisphere also develops a context-content distinction during this period. Piaget refers to the development of action schema in the eighteen to twenty-four month period as "finding" or "inventing new means." Action representation is entirely comparable to object representation.

The mental elaboration of extended pragmatic action-representations allows for a variety of action schemes to achieve the same ends. This action-context/action-content hierarchy gives rise to the conscious sense of meaning. Action schema, potential movements, contain distinct meanings. I therefore conclude that the system-consciousness synthesizes an action schema with a distinct object into a unitary mental content that can act as a new integer of consciousness for the mental apparatus.

Identity level representations also join a newly conceived subject and object. Mahler's depiction of the rapprochement child suddenly aware of aloneness in the immense universe dovetails with Piaget's cognitive description of the same age child who will

> consider himself as a mere cause and mere effect among the totality of connections he discovers. Having thus become an object

> among other objects at the very moment when he learns to conceive of their true permanence even outside all direct perception, the child ends by completely reversing his initial universe, whose moving images were centered on an activity unconscious of itself, and by transforming it into a solid universe of coordinated objects including the body itself in the capacity of an element. (Gruber & Voneche, 1977, p. 269)

Thus, the sensorimotor subjectifying infant of the first eighteen months yields the center of the stage of consciousness to a newly objectified object-self.

Neurological Development The neurological capacity for figure-ground distinction develops, according to Luria (1980), in the twelve to eighteen month period. He says that spatial orientation becomes possible

> at the end of the first year of life, when the combined work of the visual, kinaesthetic, and vestibular analyzers is established . . . Only after the establishment of this combined activity, responsible for the processes of inspection, palpation, head orientation, and eye movements, can the complex forms of reflection of spatial relationships develop, which remains unchanged even if the position of the body is altered . . . (p. 172)

The differentiation of specific object figures from the background requires both the maturation of the "feature detection apparatus," and also the use of the second signal system—speech. Here is his description of the feature detection apparatus:

> The normal process of perception of a complex visual object involves the isolation of items carrying the maximal useful information and the comparison of these items with each other. For this process to be successful, visual perception must have at least two active receptor points, one of which (macular) receives the necessary information, whereas the other (peripheral) gives information about the presence of other components of the visual field and evokes an orienting reflex, consisting of the shift of fixation, so that the object in question falls into the central (macular) area of vision. (p. 166)

Visual fixations and the formation of fixation groups that together constitute the representation of the object are the basis of both figure-background discrimination and of our sense of reality consciousness. Reality consciousness centers the identity of the young child, whereas subjective action consciousness centers the identity of the infant.

The Transformations of Language We must familiarize ourselves with the relationship between speech and language during the course of early epigenesis, for as Luria points out, all of the higher cortical functions makes use of the so-called second signal system. The development of the second signal system follows the course of epigenesis. The babbling that begins at around five or six months during the symbiotic period not only helps the development of consonant articulation and consonant hearing, but is another of the feedback activities that insures that the two hemispheres develop functional synthesis. Babbling produced by the dominant hemisphere is heard by the nondominant hemisphere, thus producing a feedback system that does not require the presence of the mother. The first or primal words have multiple referents to the object world. They are in a sense articulation of the sensorimotor transitional illusion apparatus. A single word applies to a global set of objects. During the twelve month to eighteen month period some forms of actual second system coding begin. The one year old infant may say "Hi" or "Thank you," both of which are forms that indicate the beginning or end of action programs. The twelve month old also begins to use works such as "up" or "down" that help in the coding of spatial referents as the vestibular apparatus and visual analysis are coordinated because of neurological maturation. Visual orientation requires the development of a frame of reference for the visual world that remains constant with different head positions. The twelve to eighteen month old also begins to use words like "more" that begin to show the differentiation of objects from their backgrounds. However, it is not until around the age of eighteeen months that language synthesis begins to code experience in earnest.

The postprimal scene infant undergoes a structural evolution of consciousness as the system-consciousness develops the same structure inherent in the grammar and syntax of language. Internal language becomes the intrapsychic signal system that tracks all neuropsychological processes both subjective and objective, and insures the smooth transition of neuropsychological information to intrapsychic information. The subjective neuropsychological disposition to utter speech binds language as the preferred means of expression. The development of object-constancy—the constant representational object—is paralleled by the acquisition of introjected language.

The postprimal scene cognitive revolution in consciousness gives rise to semantic and syntactic reorganizations in the integration of ego functions. Considering the dominant hemisphere changes first, the semantic function of words comes to inhibit the action system. Suspended action gives rise to the sense of meaning inherent in the morphemes of language. Morphemes, we recall, are socially agreed-upon phoneme sequences that contain the smallest or most basic unit of meaning. Thus as the child learns the semantic side of language, words or morphemes begin to inhibit action propensities. The Freudian idea that thought is trial action comes close to this description of the semantic function of language. It is important to remember, however, that according to Luria (1980) the eighteen month old child is not yet able to use speech as a perfectly reliable inhibitor of action.

> . . . at the age of 1.5–2 years a child is not yet able to subordinate his actions to an instruction spoken by an adult. The adult's spoken instruction may indeed trigger off certain movements, but it cannot make the actions performed in obedience to the spoken instructions sufficiently stable; nor in particular can it stop an activity which has already started . . . if in the course of the action another brighter or more novel object is encountered, the program evoked by the adult's instruction collapses and the child's hand is drawn to the irrelevant but brighter or more novel object. It is almost impossible to inhibit this direct orienting reaction by means of a spoken instruction. (p. 308)

Thus reality feature impressions have a powerful effect on the child of two.

In addition to the phoneme and morpheme constitution that allows words to develop into a speech system for controlling and inhibiting action, words also develop as representations of perception. The word-object becomes a concrete representative of the object: it denotes. The experience of the word as the object inhibits further perceptual feature detection of the object. In this way, inhibiting the action system and the perceptual system the language system for coding consciousness comes to form a whole representational synthesis of the neuropsychological work. The structure of language itself first derives from neuropsychological maturation, as Chomsky says, but also through its formal rules of syntactical and grammatical relationships comes to have an organizing effect on the basis of all ego functions. The sentence becomes the synthetic unity that organizes the operations of the system-consciousness in its functioning.

Luria (1980) devotes a great deal of his attention to delineating how speech mediates all of the higher cortical functions:

> *Speech plays a decisive role in the mediation of mental processes.* By being given a name, an object or its property is distinguished from its surroundings and is related to other objects or signs. (p. 31)

Thus Luria confirms the role of speech in representing reality.

A consideration of the evolution of speech after the primal scene period allows us farther insight into the operation of the young child's system-consciousness. In the period of single word utterances, generally from sixteen to nineteen months, each single utterance acts as a single synthesis, a single protosentence. Single word utterances have an action and an object sense that are simultaneously expressed. The word "mother" for instance often means "I want mother." According to Leonard (1976) who has studied the meaning inherent in single word utterances as a child's speech

> a general shift occurred from utterances related to the child's own desires, to descriptions of events the child is not necessarily involved in. (p. 21)

Thus the child's single word utterances, and unified sentences, too, move from the subject-centered to the object-centered consciousness.

III: Early Childhood: The Anal Period (24–42 months)

During this whole period of individuation and differentiation the young child's system-consciousness expands through adaptation to language. The primal scene trauma induces this prolonged adaptational process. If symbiosis is the great fixation of infancy, then the primal scene is the great trauma that ends infancy and shatters symbiosis as the interpersonal and intrapsychic mode of communication. Just as the symbiosis organized medial processes in infancy, and the sensorimotor illusion organized cognitive processes, so in this new period of life we have a medial and a lateral process. In early childhood "identification with the aggressor" accounts for medial development. Distinctions made possible through the use of language account for the lateral changes.

The autonomous child learns his or her autonomy through the development of reality energy (aggressive energy) to set up a series of self-controls. Most notable among these controls is the autonomous control over the anal sphincter, learned in part as an identification with the aggressor. The prototype of nondominant hemisphere control over the process of release of pleasure, anal sphincter control suspends the release of pleasure that occurs with the fecal discharge. As Freud notes in his *Wolf Man* case (S.E. Vol XVII), the primal scene traumatized eighteen month old child releases his feces as a protest accompanied by a scream. It is the inefficacy of the total affect storm (the temper tantrum) that triggers the adaptational transformation of system consciousness. For the essence of trauma is the inability of system-consciousness to resolve a painful situation.

The ability to deploy reality energy allows for a new form of communication with the parent. The parents are seen to be in control of the energy of reality, and the child tries to emulate the parents' control over reality in order to learn the conditions for the release of pleasure. This is the essence of "sympathetic magic," the attempt to imitate in order to gain power over the release mechanism. The sense of the medial self depends on the ability to achieve sphincter control, as well as other forms of autonomy. The sense of goodness and badness relates, in this era, to the development of self-control. The well controlled self is the one that releases discharge of all kinds in good regard to the dictates of reality. Thus, the sense of self-esteem depends on the acquisition of self-control.

The main anxiety of this era is, of course, the loss of self-control, which the child equates both with the loss of self structure and with the loss of the object's love and protection. The almost universal fantasy/fear of going down the drain, then, contains the images of the disappearance of the reality self (object-self) as well as the image of release without end and without ever achieving control. It is a fantasy of re-entering the primal scene. The fear of going down the drain therefore acts as a safeguard to the loss of the acquisition of self-possession and self-control for its outcome must be avoided at all costs. Since the acquisition of representational thinking and a strong object-self as spectator ended the original primal scene trauma, the re-emergence of such a trauma must be avoided. The fantasy/fear acts as a signal that self-control must be re-established.

During this period of autonomy development and learning about reality the parents may use pain as a conditioning agent. Spanking may begin when the child refuses to exert self-control. The pain acts as a reintroduction to trauma. Under the circumstances the child feels the self to be overwhelmed. In place of the overwhelmed medial self the child experiences him or herself as a detached object-self. The parents take on the sense of the only autonomy that matters and the child identifies him or herself with the authority of the external control. In essence this is the mechanism of the identification with the aggressor. The child's object-self is identified with the parent while the medial self is temporarily suspended in function. When the trauma is over the representation of the object-self as object is a new model for personality integration. (Kohut calls this externalized object-self, *self-object*)

The young child develops object-self and object representations based on the equation of "good" meaning in control and "bad" as out of control. All of these nondominant hemisphere processes of development are re-enforced and finally shaped through the use of language. The word "no" with its sharp, aggressive cadence comes to represent the aggressor, and at the same time comes to represent reality. In the brain/mind sense "no" comes to represent the prototypical function of the nondominant hemisphere. Language itself, we must recall, is a social signal code, and as such exists in reality as part of reality. The language that is taken in especially as command language represents the edicts of reality. The struggle for word power and word magic is the struggle for the control of salience so that reality can be commanded to produce the necessary requisites for satisfaction.

Language development during the period of young childhood follows the development of the system-consciousness. The sentence represents a unification of conscious processes, and the structure of the sentence itself comes to represent the four zones of consciousness. In this way the sentence accounts for each zone of tertiary consciousness in its own synthesis. At the beginning of the period of early childhood the sensorimotor subject-centered consciousness is still in vogue. Gradually the child learns to distinguish a subject "I" from an agent "want." In linguistic terms, the agent is the predicate of the sentence, i.e., the action verb. As verbalization goes on and develops, the important part of the sentence shifts to the object "mother."

Clearly, the function of naming is an important differentiating mechanism for system-consciousness. The neuropsychological ability to designate or denote objects through naming is an aspect of the nondominant hemisphere function of representing percepts. Because spatial auditory zones mature in this era, sound can be localized as existing within objects. The naming function as we can see with onomatopoeia is based on the neuropsychological disposition to localize sound as a portion of reality within objects. Thus the intonations and cadences of languages, its prosodic rhythm, remain close to the reality of the object.

During the two to three year period the possessive *my* begins to come into use a great deal. This reflects the development of the object-self as an object in the world. The ability to turn the sentence around so that the child sees action as centering in the object shows a further advance both in language and in system-consciousness. Thus "mother put" shows the young child taking the command role as if he represented reality and is addressing the subjectivity in the object. The ability to perceive and suspend action emanating from the object world means that other persons are truly seen as autonomous. Language development helps in all of these distinctions.

IV: Oedipal Myelinization and Symbol Formation: Childhood (4–7 years)

Overview Between four and seven the posterior cortex myelinates in the parietal "supra modal" areas. At the same time the anterior cortex myelinates in the premotor and prefrontal areas. This expansion of functional cortex potentiates and then ensures a total revision of the distributed system-consciousness. The process involves three phases: (1) the transformation in the system-consciousness that occurs around the age of four, (2) the system-consciousness ontology during the years of continuing fast myelinization, and (3) the end stage in which a final transformation—reflective synthesis—supersedes the transformations of the oedipal period. The final stage lays the ground for latency.

While the neurological developments of the oedipal period put a ceiling on the drive-organized subsystems of consciousness through the development of newly functional action inhibition, they also make a floor for consciousness's new volitional system. Until the age of four, motivation and response were still determined by instinct. After the age of four, system-consciousness is also regulated by an interaction between its own voluntary mode and the social forces it cognizes in accordance with its own organization. The prefrontal cortex that myelinates during this era is known by Luria as "social cortex". Its functioning supports the hierarchically superordinate "higher cortical functions." As the social cortex functionalizes, the child feels a *compel-*

ling motive force in addition to the *impelling* motive force of the drives.

Luria and Vygotsky (1980) formulated a principle that states the regulatory significance of the higher cortical functions when cortical damage occurs:

> In the early stages of ontogenesis, a lesion of a particular area of the cerebral cortex will predominantly affect a higher (i.e., developmentally dependent on it) center than that where a lesion is situated, whereas in the stage of fully formed functional systems, a lesion of the same area of cortex will predominantly affect a lower center (i.e., regulated by it). (p. 35)

This rule applies to intrapsychic process as well. Pre-oedipal-age traumas and fixations, as well as brain damage, produce developmental disturbances of system-consciousness organization. Oedipal or post-oedipal traumas and fixations produce intrapsychic conflict.

The Neurological Origins of Oedipal Fantasy The revolution in system-consciousness that eventuates in a voluntary speech system using symbols as the medium of intrapsychic consciousness is triggered by *revised primal scene experience.* Childhood sexuality effloresces as phallic urethral masturbatory impulses reach a peak necessity for consummation around the fourth birthday. As the consummatory wish to possess the object of desire in a total, though not yet fathomed, way increases, and as the accompanying desire to remove all obstacles to fulfillment of masturbatory release increases, there is a readiness to experience an exclusive encounter between the parents as a sexual trauma and fixation. The completely facilitated wish and fear of laying into the midst of the parental embrace engenders fantasies in the child that are universal: because they are the biological response to the newly myelinizing cortex now ready to be kindled to functionality.

The fantasies engendered in this period become the guardians of the neurological sanctity of the medial drive channels. Without new fantasy to channel the increased oedipal drive pressures for discharge into new experiential channels of newly readied cortex, old channels would be insufficient to provide the experiential means of gratificatory possibility. Fantasies provide the means for system consciousness to organize the drive processes in such a way that the problem solving cognitive modes of system-consciousness can relieve the drive pressure. A loss of impulse control in the oedipal period, with bed-wetting or fecal discharge, is one form of unregulated discharge of masturbatory consummatory drive. Affect discharge (consisting of new forms of tantrum behavior or of regressive discharging tantrum behavior with lack of behavioral controls) signals both failure of fantasy modes to control the disposition of drive energy, and the gradual channeling of drive pressure into new medial fantasy discharge patterns.

The formation of new *universal fantasies* occurs in a biologically mandated sequence. The transformation of early childhood occurs in waves of new experience induced by the limbic system. Revised primal scene episodes of jealousy of each parent give rise to the positive and negative versions of the oedipal fantasies. These all or nothing fantasies that rechannel the drives aim for (1) complete consummatory gratification of the subject-self through possession of and fusion with the object of desire, and (2) destruction of the interfering object who is in possession of the desired source of all gratification. The all-or-nothing, complete action and reality drive mobilization by the revised oedipal wishes is the hallmark of revolutionary change in system-consciousness, for such a mobilization announces that the system-consciousness of the preceding anal period is no longer capable of mediating the drive aims.

The new experience that is driven by mobilized biological forces and directed by new universal fantasies recruits the newly functional cortex to transform the identity functions. Thus, the newly functional inferior parietal cortex coalesces a new object-self body image, while the newly functional superior parietal cortex allows for the formation of new perceptual object entities. The self function regulates the new urethral discharge possibilities. Alternating anal and urethral sensations of discharge produce the capability for orgasmic experience. The subject function integrates these masturbatory pressures for consummation. The agent function integrates a new form of action integration that is future oriented, an action integration on which the functional imagination is based.

In this era the working relationship among the identity functions is determined by the new aims of system-consciousness. I have said that the aim of both universal oedipal fantasies is to reduce the intrapsychic distance between subject and self in order to find the complete masturbatory pleasure of discharging the working energy of the system-consciousness. This wish becomes the aim of the system-consciousness in this period. With the aid of a new form of thought, *symbolic thought,* the oedipal child proceeds to employ a newly formed speech system, a voluntary speech system, in the attempt to bring about the reduction in separation between subject and self. Oedipal subjective consciousness functionalizes as a realm of imagination in which symbolic play activity expresses the egocentric world. The derivatives of the universal oedipal fantasies organize the new egocentric world of imagination.

Imaginative play is the behavioral accompaniment of egocentric speech. It provides a framework for expanding the repertoire of symbols. Late in the oedipal period, toward latency, egocentric speech achieves formal reflective organization as *internal narration.* It branches into two speech systems: inner speech and social speech. Egocentric inner speech, the basis for functional voluntarism, is organized in the dominant hemisphere. The ontology of this speech system accounts for the higher ego function of planning. When imaginative action programs of the oedipal period are blocked by reality, the child learns to address an adult in problem solving social language. This form of address is internalized as social speech. Gradually a social speech system develops, materially re-enforced by the social identifications that set in as latency approaches.

The oedipal child gradually develops a split in identity consciousness between the subjective, fantasy-filled action child, and the social child who belongs to the family and the community. When, at the end of this period, oedipal fantasies are shattered—exposed as illusions—the major emphasis of identity-organizing conscious processes shifts to the nondominant hemisphere. This shift is accompanied by the sense of consciousness as an objectified reflective identity entity. The latency child is ready to be imbued with the objectified thought patterns and language forms of the culture. The oedipal child therefore moves from a system-consciousness that has been medialized and subjectively centered by efflorescing action pressures (around the age of four) to a new system of consciousness that (by the age of seven) is lateralized and objectified by reality hemisphere forces. By the seventh year, right hemisphere processes exert a major influence, commanding the preponderance of EEG organizing potential when the child is called upon to think. This is a situation that remains stable until preadolescence, when action forces effloresce again, this time under the influence of hormonal changes. In adolescence, the increased action-drive pressure triggers the next major transformation and new hierarchical development of consciousness.

Ego Function Development in the Oedipal Period Luria provides a background context for understanding the integration and synthesis of ego functions and their ontology in his book *Higher Cortical Functions in Man.* According to his framework higher intellectual operations require the formation of hypotheses that are developed on the basis of plans that require perceptual validation by salient features that have heuristic value if the hypothesis is correct. The action program, or plan, that is developed to test the hypothesis is formulated in terms compatible with the operation of the dominant hemisphere based inner speech system. By adulthood, the relevant aspect of inner speech consists of a formal series of propositions that require a series of essential operations to be performed in order to test the hypothesis. Just as inner speech has contracted to form essential propositions, so has social speech contracted to provide a context that contains all of the features that must be present to prove out the hypothesis. This salience context is prepared in intimate collaboration with nondominant hemisphere prefrontal lobe functioning.

Luria (1980) describes the neuronal area maturation that makes the progress toward these higher operational developments in the four to seven year old child:

> While the neuronal apparatuses of Area 4 have attained full development in a child approximately 4 years of age, Area 6 continues to develop and reaches full maturity only in a 7 year old child. At this time the premotor region of the cortex also greatly exceeds the motor region in area. (p. 217)

Area 4 is the motor area and area 6 is premotor. The development of the premotor cortex is necessary to the development of inner speech because it provides for the selective suspension of action which is essential to the formation of pure thought. Luria (1980) cites research relevant to the change in the 4 year old's method of processing speech:

> . . .when or shortly before the child begins to attend school the hearing of speech ceases to require the active participation of articulation. However, if the child is told a word with a complicated sound, or still more, asked to write it, the articulatory apparatus will again be brought into visible use to aid in the perception and recognition of the precise sound structure of the word. (p. 115)

During the preceding representational period the premotor cortex inhibits irrelevant sounds via articulation sequencing.

In the oedipal period expansion of the premotor cortex allows a hierarchical sequencing function to take over the motor dynamics of phrase making. We can get a practical sense of premotor sequencing when we think about the inner blanks and their rhythm that we fill with speech as we make a sentence. Certain words and phrases in the language have evolved to include this motor tendency. Phrases such as "so on and so forth" or "etcetera, etcetera, etcetera" are examples.

Luria (1980) asserts that the fluency of internal speech is entirely dependent on the integrity of premotor functioning:

> . . . the kinetic schemes of a language are more than words. Jackson (1884) expressed the opinion that the phrase, not the word, is the unit of speech ("propositional speech"). Similar views have been put forward in modern linguistics (Chomsky, 1957) in relation to the whole expression. Every expression possesses some form of dynamic structure with the individual parts requiring specific turns of speech for the phrase to be completed. By means of these dynamic structures, determined at the beginning of each expression, it is possible to foretell fairly reliably the manner in which the expression will continue. (p. 230)

The development of the kinetic dynamics of phrase making is on a higher level than the kinetic dynamics of simple phonemic succession. The period of rapid expansion of the premotor cortex beginning at age 4 makes inner speech possible. He goes on to point out that this ability to relate action to a motivated program connected to communication is a frontal lobe function. The myelinization of the prefrontal cortex, as opposed to the myelinization of the premotor cortex, allows motive forces developing out of limbic sources to integrate with action plans, which develop out of lateral convex cortical sources. Premotor maturation is necessary for the sequencing of action and speech dynamics while the maturation of the prefrontal cortex is necessary for the inclusion of motive in the action plans and speech plans conceived.

The prefrontal cortical development allows subject function integration to be smoothly meshed with agent function action planning. According to Luria (1980),

> The complexity of the neuronal structure of the fields of the prefrontal region is confirmed by the fact that they develop much later in ontogenesis. Flechsig (1920), the first to use the myelogenetic method, showed that the fibers of these cortical divisions are the last to myelinize and that this region of the cortex begins to function later than the other regions. (p. 259)

Thus, the prefrontal region is hierarchically pre-eminent as synthesizing cortex.

Symbol Formation In this period the ego-function *symbolic-thought* becomes the synthesizing function of system-consciousness, just as *representational-thought* was the synthetic function during the previous period. Symbols are integers of hierarchical generalization, grouping information in all four zones of integration of consciousness. The higher the level of generalization, the greater the ease of intrapsychic communication in the manipulation and synthesis of the information derived from brain operations. Both the dynamic action groups of the dominant hemisphere premotor functioning, and the selective perceptual grouping of the nondominant hemisphere are based on the ability to "repress" (selectively inhibit) whole categories of information, and to selectively facilitate other categories of information that are symbolized by particular features. The symbol becomes an integer of consciousness both for what is included and what is excluded from consciousness. This new layering of system-consciousness increases intrapsychic efficiency.

Symbol formation also allows for a restructuring of emotional organization. In every person's childhood development, certain ubiquitous objects take on childhood significance. The breast, the penis, the baby, fecal matter, and the stream of urine are main examples. These body-identified objects have a representational significance in the representational period; they had a transitional significance as unifying objects in the symbiotic period of infancy, and they acquire a symbolic significance in the oedipal period. Symbols, we see, contain a whole ontology of highly condensed developments of consciousness. They therefore act to integrate and synthesize it.

Vygotsky (1962) shows how egocentric speech begins as an accompaniment of physical activity, develops into an approach to problem-solving and finally becomes inner speech. In his example of a five and one-half year old boy's drawing of a streetcar we see the linking of egocentric speech to activities that express the child's consummatory wishes and motives. The child's pencil broke:

> He tried, nevertheless to finish the circle on a wheel, pressing down on the pencil very hard, but nothing showed on the paper except a deep colorless line. The child muttered to himself, "It's broken," put aside the pencil, took watercolors instead, and began drawing a *broken* streetcar after an accident, continuing to talk to himself from time to time about the change in his picture. (p. 17)

Here, intimacy with subjective motives gradually becomes a part of the developmentally more advanced ego function of planning.

In successive experiments, Vygotsky (1962) found a developmental order in the use of egocentric speech. It appeared first at the end of an activity, or, as in the streetcar observation, it marked a shift in the child's perception of his own activity. Later, egocentric speech appeared at the middle of an activity, and still later it shifted toward the beginning,

> . . . taking on a directing, planning function and rasing the child's acts to the level of purposeful behavior. (p. 17)

Thus, during the period from about four to seven egocentric speech develops from an accompaniment of action ensemble, to symbol of the organized action, to precursor symbol, useful in the planning of the action.

Vygotsky studied the syntax of inner speech. He found that it is a contracted form of speech in which the subject is left out entirely and in which the predicate appears in a foreshortened form:

> . . . as egocentric speech develops it shows a tendency toward an altogether specific form of abbreviation: namely omitting the subject of a sentence and all words connected with it while preserving the predicate. This tendency toward predication appears in all our experiments with such regularity that we must assume it to be the basic syntactic form of inner speech. (p. 139)

This most intimate form of subjective expression leaves out the words of the subject. The subject of inner speech is so deeply implicit it needs no inner conveyance. Vygotsky elaborates, pointing out that in the most intimate conversations such as the conversations of love, where the participants are deeply in tune with the mutual access of their feelings that inner speech is a sufficient prototype for the most momentous exchange. Although it expands to propositional function, inner speech remains in contact with the intimate consummation motives of the individual.

Vygotsky (1962) describes the development of the phenomenon of predication in inner speech:

> The predominance of predication is a product of development. In the beginning, egocentric speech is identical in structure with social speech, but in the process of its transformation into inner speech it gradually becomes less complete and coherent as

> it becomes governed by an almost entirely predicative syntax. Experiments show clearly how and why the new syntax takes hold. The child talks about the things he sees or hears or does at a given moment. As a result he tends to leave out the subject and all words connected with it, condensing his speech more and more until only predicates are left. (p. 145)

Predication gradually gives rise to the dominant hemisphere semantic function.

Vygotsky (1978) points out that as egocentric speech becomes interiorized and finally contracts to inner speech, another branching development also occurs. When the child is blocked in his play he increasingly turns to the adult for help in continuing or repairing an action plan. In order to get help the child must use a more complete form of communication than inner speech. Therefore the child learns to tell what difficulty he is encountering. This telling, Vygotsky states, develops into the function of *social speech*.

> . . . egocentric speech is linked to children's social speech by many transitional forms . . . the link between these two language functions occurs when children find that they are unable to solve a problem by themselves. They then turn to an adult, and verbally describe the method that they cannot carry out by themselves. The greatest change in children's capacity to use language as a problem solving tool takes place somewhat later in their development, when socialized speech (which had previously been used to address an adult) *is turned inward* . . .
>
> . . . the process of the internalization of social speech is also the history of the socialization of the child's practical intellect. (p. 27)

Vygotsky's description of the process through which inner speech becomes socialized corresponds to the psychoanalytic mechanism of *internalization.* Internalization is a mechanism that develops in the five year old period. Throughout epigenesis we have seen that assimilative mechanisms account for the subjective encompassment of information sets, and other, accommodative mechanisms account for the objective accumulation of information.

Internalization is a mechanism through which inner speech encompasses action patterns in reality. Gradually these action patterns can be initiated and carried out as action plans. By contrast, the mechanism of identification operates through an imitation of such patterns of reality inhibitions and regulations that the object-self gradually becomes more and more like the object. At the end of the oedipal period identification is the pre-eminent mechanism for producing further change in system-consciousness.

There is a gradual shift during the ontology of the oedipal period from egocentric to narcissistic-inner modes and finally to object-directed social modes of relating to the world. Gradually the changes in consciousness that are accomplished through internalization of the adult action modes are solidified through identification mechanisms. This shift is consolidated by the maturation of a new hierarchical form of anxiety: castration anxiety.

The symbols that emerge in this period are experienced as extensions of the drive aims. Bringing the drive aim into consciousness as organizers of behavior, they weave into the universal organizing fantasies that pattern psychic life and thus allow the derivatives of these fantasies to become conscious. Symbols bring the imagination into the formation of identity goals, for they are felt to contain within them the omnipotential of gratification. The various body symbols coalesce in the constitution of the self as an affective symbolic entity and as a component of identity. The affectively reverberant symbols achieve object-self identity signification as well. Thus, the aim of the drive is symbolically signified as possession of or by the *baby, penis, breast, fecal column,* or *urinary stream.* Eroticizing the body parts and products makes them symbols of the whole consummation aim. As identity integer, object-self coalesces in a new symbolic way that subsumes the previous representational object-self. In this era a part of the object-self can represent the whole.

Each symbol is capable of symbolizing various identity functions. Consider the symbolism of "the baby." Sometimes the child imagines itself as a baby. While *baby* clearly symbolizes the integrity of the body of the past—and of every other child's past as well—this symbol also remains available to represent the child's self-image of his or her present body, for the teddy bear, the doll, the favorite creatures are object-self symbols at the same time that they represent the self as a baby in the present. *Baby* also symbolizes the self function of the inner body, for when asked, the oedipal child says that his or her self is a smaller body located at some depth beneath the skin. More than this, *baby* symbolizes the subject function of pleasurable consummation, for the image of the baby in the womb is often experienced as the aim of childhood masturbation. The correlation of *baby* with sexual gratification often continues into adult life, so that one often hears a grown person use "oh baby!" to announce the imminence of pleasure. The symbol is therefore an organizer of consciousness of this era, and the crossroads of identity functions.

The Oedipal Expansion of System-Consciousness Vygotsky described the function of play in expanding the consciousness of the oedipal child. Play makes use of symbolic thought to form elaborated fantasies that combine the underlying universal fantasy motifs with new action and perceptual possibilities. Play is a highly subjective kind of activity in which the emphasis is on wishful initiation and on action sequences that are meant to show the wishes actuated. Imagination is the continuation of this activity. The egocentric speech that accompanies play is also a highly subjective, dominant hemisphere centered kind of ego function.

As Vygotsky (1978) says:

> Toward the beginning of preschool age, when desires that cannot be immediately gratified or forgotten make their appearance and the tendency to immediate fulfillment of desires, characteristic of the preceding stage, is retained, the child's behavior changes. To resolve this tension, the preschool child enters an imaginary, illusory world in which the unrealizable desires can be realized, and this world is what we call play. Imagination is a new psychological process for the child; . . . Like all functions of consciousness, it originally arises from action . . . (pp. 93–94)

The early oedipal period is extremely egocentric and subjective.

Vygotsky describes the ontology of egocentric speech as the oedipal child moves along toward a middle, more narcissistic, period of oedipal development.

> Our experimental results indicate that the function of egocentric speech is similar to that of inner speech: It does not merely accompany the child's activity; it serves mental orientation, conscious understanding; it helps in overcoming difficulties; it is speech of oneself, intimately and usefully connected with the child's thinking. Its fate is very different from that described by Piaget. Egocentric speech develops along a rising, not a declining curve; it goes through an evolution, not an involution. In the end, it becomes inner speech. (p. 133)

As Luria (1980) amplifies:

> . . . "internal speech" is of fundamental importance in the formation of the integral system of the sentence. According to Vygotsky (1934, 1956) it is internal speech, with its predicative properties, that performs these dynamic functions and plays a direct part in both the evolution of the thought into the whole expression and the crystallization of the whole expression into its shortened, conceptual scheme . . .
>
> Every word can therefore be regarded not only as the bearer of meaning of a particular object, but also as the unit of expression, with potential connections that unfold as the word is included in the complete sentence. (pp. 230–231)

The agent function contains the inherent identity integration for this set of premotor based ego functions. We may think of voluntary identity as an expression of agency. The agent function integrates inner speech. It is during the oedipal period that agency takes on its operational characteristics. The phraseology of an intention comes to compose a unit of thought. We may think of the agent wielding action verbs in a sequence that outlines an action process under consideration.

Luria points out that action is evoked in early childhood through orienting stimuli, later by commands, and finally in the oedipal period by voluntary processes capable of inhibiting unnecessary movements. The transition to the ability to inhibit action through the formulation of internal speech does not begin until the age of four, and then it gradually increases in scope. This corresponds to the transformation in consciousness ordained by the rapid myelinization that begins around age four.

> . . . Not until the age of 3.5–4 years does the action program evoked in the child by a spoken instruction become strong enough to ensure the necessary action without distraction and to inhibit irrelevant activities . . . *The ability to subordinate one's action to a program formulated in a spoken instruction and to inhibit more elementary forms of actions arising in response to direct impressions or a result of perseveration of previous actions is formed only gradually and is the product of prolonged development.* (p. 309)

"Castration anxiety" has two forms, a dominant hemisphere form, and a nondominant hemisphere form. The dominant hemisphere form operates to prevent mistaken actions. The child begins to use this form of action-anxiety to prevent the execution of imaginative action plans that would result in violations of the child's real world. He must not stab a playmate with a knife while enacting an oedipal drama. The action-anxiety mechanism of this era requires the child to distinguish egocentric and imaginative plans from social plans. The discrimination between fantasy play and social conduct is, of course, the action testing component of "reality testing." The result of failed play is a sense of damaged subject-esteem with feelings of decreased confidence in play, and inadequacy.

The nondominant hemisphere form of "castration anxiety" is based on preventing damage to the body. The possibility of real damage to the body, or even of real castration, is a symbolic realization that inhibits the child from attempting to realize oedipal fantasies outside the realm of symbolic play. Thus reality testing distinguishes the real world from the fantasy world. Oedipal fantasy derivatives are useful in helping the child to explore the various roles that are possible in reality. Indeed, the oedipal child becomes intensely interested in differentiating secondary sexual characteristics. Secondary sexual features, including clothing, structure role differentiation in this period. However, insofar as these secondary sexual features also begin to define the object-self and the object-self's anxiety about damage to the body or to the potential acquisition of the secondary sexual features, they operate through the identity zone of reality consciousness. The fantasy aim of destroying the sexual rival gives rise to an uncanny and severe form of nondominant hemisphere castration anxiety, for the talion principle operates through the mechanism of object-self identification with object. Destruction of either parental object is tantamount, even in fantasy, to object-self destruction.

The objectification of "the mind" is not firmly established within an identity framework until the end of the oedipal period. During the oedipal period thoughts are still not localized within an identity integrating object-self. Thus the oedipal child feels that his destructive oedipal fantasies invite immediate retaliation as his parents must know about them. The oedipal child is just learning that it is possible to lie. He still believes that "Santa Claus is coming" and that 'He knows when you've been good or bad."

The Shattering of the Oedipal Fantasies Both the cognitive and the emotional organization of oedipal life move toward a shattering in the sixth year. Reality testing based on castration forms of anxiety places more and more limits on the oedipal child's mode of fantasy operation. The internalization process through which the adult world is assimilated to the play routines of inner speech finally gives way. Increasingly, the peer group insists on play that is social and that abides by rules. At first rules are just internalized generalizations of action possibilities. However they increasingly become moral edicts, inviolable givens. The force of reality increases at the end of the oedipal period. Just as the primal sensorimotor subjective period shatters under the impact of the primal scene, giving way to reality enforced commands and speech patterns, so also does the oedipal period egocentric fantasy mode shatter under the impact of the weight of reality, the danger of the fantasies, and the neurological maturation that beckons between six and seven.

Latency-shaping identifications consist of massive adaptive changes in the personality structure. At the advent of latency the nondominant hemisphere once again comes into the ascendency. Around the age of seven, neuropsychological maturation of the prefrontal cortex forces socialization. Luria (1980) has discussed the shift to right hemisphere functioning that occurs at this age. By channeling different words into the left and right ear simultaneously (dichotic listening) it is possible to test differential recall and therefore dominance of function for language hearing. Thus:

> . . . What is particularly interesting is that in children of 6–8 years a "negative right ear effect" can be seen much more often than in adults: in other words, the range of recall of words presented to the left ear is greater than to the right. The frequency with which this phenomenon is found decreases with age: in children of 6–8 years a "negative right ear effect" is observed more often than in children of 10–12 years. In adults it is found in about 5% of cases. (p. 100)

Luria backs up his contentions about the pre-eminence of the right hemisphere for various functions beginning at age six to eight with clinical evidence from brain lesions in children:

> . . . the mechanisms lying at the basis of hemispheric dominance in children are different from those in adults . . . the bilateral effect characteristic of lesions in the dominant hemisphere arises in children with pathological foci not in the left, but in the right hemisphere. . .
>
> . . . in childhood a disturbance of function arising through a lesion of the left hemisphere is very likely to prove compensatable by activation of the opposite, right hemisphere. . .
>
> . . . symptoms due to a lesion of the right hemisphere . . . as spatial agnosia, unawareness of the left side, disturbances of drawing, and apraxia of dressing, evidently connected

> with disturbances of the body schema, are manifested much more clearly in childhood than the symptoms of a lesion of the left hemisphere. . . (p. 101)

The right hemisphere determines identity awareness in latency.

The Formation of the Superego and the Ego-Ideal A revolution in consciousness takes place around the sixth year. Ending the oedipal period, it ushers in *latency*. Freud calls this revolution the "shattering of the oedipus complex." It makes functional a reflective, self-observing level of system-consciousness. The new consciousness is hierarchically superior to that of the oedipal child. Internalized action instructions become the foundation of the social speech system and give rise to the ego-ideal. An ability to identify with rules and regulations becomes the basis of superego.

To put the case developmentally, the shattering of the oedipus complex redistributes drive energies and so produces a new set of identity-functions. A reorientation of the drives appears in the wake of the defeat of the oedipal fantasies which are exposed as mere illusions. Between the ages of five and six the belief in the reality of retaliatory castration becomes so complete that the child must put aside his wishes to annihilate his rival. Freud has pointed out that the simultaneous shattering of positive and negative oedipal fantasies results in a new formation of massive internalizations and identifications of mother and father composited as ego-ideal and as superego. He theorizes that the liberated libidinal energy (action energy) becomes available for the formation of new sublimations and ideals which make up the ego-ideal, and that the liberated aggressive energy (reality energy) becomes available for the development of moral values which are the imperative of the superego.

The distinction between ideals and morals corresponds to the distinction between dominant hemisphere organized subjective processes, and nondominant hemisphere organized objective processes. Shame (mediated by the dominant hemisphere) and guilt (mediated by the nondominant hemisphere) regulate the drive forces. Shame is a subjective regulation in which consummation strivings and expressions fail to meet ideals. Shame, as humiliation, occurs when action-performance fails to conform to ideal standards. This depletes narcissism. Guilt has to do with objective regulation. Here the self and object-self are felt to deny moral injunctions and social rules. The social object-self acquires inhibitory command at the expense of the affective self.

Forming intentions (making plans) and observing rules, the newly functional portion of system-consciousness produces a reflective mental synthesis. As a result, the capacity for continuous self-observation develops. This superego function protects the stability of the new object-self. The child feels himself to have become a stable participating and social member of a peer group, which in turn is a branch of society that includes the parents.

In the wake of the shattered oedipus complex, every object relationship must be experienced as an ideal or moral relationship. The unification of identifications and internalizations produces a social fabric that provides the experiential basis for the new layer of abstract consciousness. The parental figures tend to recede into a social fabric in which they are the main designs, but in which they no longer comprise the whole of reality. The world has been abstracted to their likeness and generalized to the new social cortex which responds to a variety of experiences in similar ways. Thus, after the shattering of the oedipus complex, new experience is profoundly connected to the social nature of the world.

V: Latency (Age 6–11)

Like the two to four year old representational period, latency is reality centered. Like the earlier period, latency begins in a traumatic shattering of subject-centered illusions. The shattering of the oedipal world of imaginary play is accomplished through an adaptational process. Latency's problem-solving stability is the prototype for adult problem-solving stability. During latency system-consciousness develops a layered capacity to integrate symbols that results in the formation of the capability for abstract thinking.

The cognitive advances of latency result in a form of thinking that Piaget calls "operational thinking." This is a form of problem-solving based on the ability to understand physical cause operating in an objective realm of reality. For example, this is the period in which the child is able to suspend a subjective belief that the moon is following him or her in favor of an explanation centered in physical reality.

The emotional system advances of latency are based on the reflective layering of the subject and self identity functions. The relative completion of prefrontal myelinization around age seven allows a hierarchical form of regulation of both consummation strivings and discharge processes. The internalization process of the oedipal period results in a massive ideal system formation in latency. Latency ideals correspond to a system of justice in which action programs are circumscribed according to idealized versions of the parents' acting in heroic and unselfish ways. The social ideals put a lid on consummation strivings. Similarly, the identification processes that follow in the wake of oedipal shattering form a layering of self and object-self identity functions that promote moral rules as the representation of reality principles.

The system-consciousness of latency forms a stable reflective layer of identity functions. This means that self-observation becomes a continuous process. Latency is the period of life when long term memory becomes continuous. The reality centered focus of identity functions makes objective self-observation into the prototypical form of new consciousness for this era.

Reflective Abstraction: Fields of Time and Space The abstract layering of system-consciousness that develops in latency brings a new coherence to cognitive functioning. The dominant hemisphere mode of sequential processing gives rise to a new form of action plan that is consciously sequenced in terms of past, present and future. The sense of personal subjective time becomes a reliable subjective identity function. The nondominant hemisphere mode of simultaneous integration gives rise to a new abstract sense of space that is consciously perceived as infinitely extendable. The emphasis on the nondominant hemisphere functioning in latency can be seen in the child's capacity to tell time. The child learns to perceive time as an abstract and objective phenomenon, perceivable as spatial displacements on a clock. This shifts a subjective function into an objectified form of abstraction.

Vygotsky formulated the manner in which the latency age child uses symbols in the constitution of a *uniform dynamic field of attention,* which replaces both the immediacy of perceptual field observation and the action system's bondage to stimuli. He points out that a new attentional state created by the child replaces the perceptual fields, and that, in becoming aware of the succession of such attention fields, the child creates his sense of subjective time.

Thus, the dominant hemisphere proposes an action state or verbal proposition formulated in signs, while the nondominant hemisphere looks for signifiers in reality instead of taking the whole field of reality as its perceptual object. A new reflective synthesis of the attention states produces a dynamic field of attention out of which a sentence or an action may emerge. As Vygotsky (1978) puts it:

> With the help of the indicative function of words, the child begins to master his attention, creating new structural centers in the perceived situation . . .
>
> In addition to reorganizing the visual-spatial field, the child, with help of speech, creates a time field that is just as perceptible and real to him as the visual one. (pp. 35–36)

Thus the attentional field combines elements of perception (reality sense) with elements of action process in a new synthetic mental structure.

For problem-solving the individual must *propose* a series of actions. These must be perceptually confirmed in order to solve the problem at hand. Such propositions are the work of the dominant prefrontal tertiary cortex, as well as of the medial prefrontal tertiary cortex. Luria points out that propositional speech is an outgrowth of inner speech with its predicative drafting of potential solutions in action words to problems. Vygotsky proposes that the attentional fields form a sequence which corresponds to the propositions in a problem solving series:

> The transition from the simultaneous structure of the visual field to the successive structure of the dynamic field of attention is achieved through the reconstruction of the separate activities that are part of the required operations.
>
> . . . Created with the help of speech, the time field for action extends both forward and backward. Future activity that can be included in an ongoing activity is represented by signs . . . This emerging psychological system in the child now encompasses two new functions: *intentions and symbolic representations of purposeful action.* (pp. 36–37)

These new twin functions, synthesized in the dynamic field of attention in the latency age child correspond to dominant and nondominant hemisphere major ego functions respectively. They are the cognitively graded lateral functions corresponding to the medial object related social functions of ideal formation and moral formation. It therefore becomes possible to perceive that reflection in latency integrates the more lateral or convex functions with the more medial. On the dominant side we discern intentionality, which becomes less cognitive as the intentionality comes into conjunction with ideals, and so regulates the degree of drive discharge potential inherent in the intention. On the nondominant side we can discern the abstract representation of the completed act in its perceptual context, which must be regulated by the rules and regulations which enforce the maintenance of a stable reality. In other words, the abstractly conceived end point of the action must be weighed against the moral scruples which control voluntary action.

Vygotsky (1978) directed himself to this point. He said that during the reformulation of the structure of consciousness there were also changes in the child's motivational system which made the motive force more social. Following K. Lewin, Vygotsky asserts that the child's motives become socialized quasi-needs, secondary to the solution of the problem at hand. This gives us an insight into the aim of consciousness in the social period of latency. The aim is problem solving: providing a match between propositions generated with the introduction of a problem and socially appropriate solutions to the problems.

> . . . with the development of these quasi-needs, the child's emotional thrust is shifted *from a preoccupation with the outcome to the nature of the solution* . . .
>
> Because he is able to form quasi-needs, the child is capable of breaking the operation into its separate parts, each of which becomes an independent problem that he formulates for himself with the help of speech. (p. 37)

VI: Neurohormonal Transformations of Adolescence (Age 11–21)

Each phase of increase in action resources produces a subjective bias in the hierarchy of functions. A movement toward subjectification is always followed first by a reassertion of reality-oriented functions, and then by a new equilibrium that allows stable developmental change to accrue over a period of time. To give an example, the primary symbiotic sensorimotor world of transitional creation leads to a period of stabilizing autonomous, reality-centered development. Similarly, the subjective period of oedipal egocentrism yields to the latency age stabilization which is induced by a vast prefrontal myelinization and which is maintained by the *concrete mental operations* of latency system-consciousness. Neurohormonally induced, adolescence refocuses consciousness again.

Adolescent thinking is characterized by the development of *formal thought.* This is a cognitive process in which, as Piaget says, the adolescent superimposes a spectrum of hypothetical possibilities on whatever happens to exist in reality (Inhelder & Piaget, 1958). Hypothesis making, then, rather than integrating the givens of reality, is the major function of adolescent system-consciousness. This means that the development of formal adolescent thought replaces accommodative reality cognition of latency with a new form of egocentric assimilation. The imaginative, libidinized, and subjectified hypotheses of adolescence overwhelm latency's stable operations until the advent of adult consciousness shifts the balance back toward the nondominant hemisphere again. The adolescence-long transformation of latency-age operational thinking is a prolonged growth process of identity transformation that begins in puberty and continues up the threshold of adult life around the age of twenty-one. Adolescence progresses through three marked phases of development: the *egocentric,* the *narcissistic,* and the *object-related.* These phases correspond to the identity forming stages that followed one another in earlier mind/brain maturation. The hormonal maturation which initiates adolescent identity changes proceeds at a chronologically slower pace than it did earlier. New identity transformation takes a full decade. Higher levels of growth and sex hormones trigger adolescence's highly subjective egocentric stage. These increases must augment the scope of the dopamine action-drive. It is marked by new highly subjectified action processes. The peer group (all affected by the same pressures) resembles the peer group in the early oedipal period when children engage in parallel egocentric play, which is to say that they all engage in newly initiated action processes and they all fixate to genitally matured masturbation. This process subjectifies other people. The peer group is seen as an extension of the derivates of subjectively activated masturbation fantasies.

A movement away from oversubjectification characterizes the narcissistic phase of adolescent identity transformation. Mid-adolescence is characterized by a strong synthetic drive which brings the intensified subjective love into contact with a world that is still conceived as an extension of the adolescent's own purpose. A deep well of love is expended on externalized object-self and spills over to the whole world conceived as an extended object-self. This narcissistic form of love is, in essence, the process of transforming ego cathexis ("I") to object cathexis ("me"). The love of the object-self as object is the form of relating that Freud cites in his paper *On Narcissism* as the way station to complete object love. The increase of sex hormones during this phase neuromodulates intensified love processes.

The further shift to objectification and to reality centered processes appears to be neuromodulated by the augmentation of the thyroid system. Thyroid-stimulating hormone enhances the capacity of the whole system-consciousness for problem solving. As the hormonal systems mature one by one in adolescence the whole capacity of system-consciousness increases until at the onset of adult life a highly stable and homeostatically stabilized form of problem solving system has evolved.

The hierarchical coalescence of the new system of consciousness is manifested in the development of *meta-perspective.* The symbolic inner speech system of the oedipal child, and the abstracted social speech system of the latency child are generalized to conceptual levels in adolescence. The experience of identity functions is then elevated to corresponding conceptual levels. The final synthesis of identity in the late adolescent identity-crisis entrance into adult psychological life consists of the formulation of an objective construction of personal identity—a personal theory of mind—that the young adult carries into life as his or her own metaperspective.

Neurophysiological and Neuropsychological Developments in Adolescence Radically changed hormone levels in all three neural systems fuel the identity change progression in adolescence. I theorize that each neural system—the neutral, the action, and the reality—has extended hormonal components that feed back and increase the drive capacity of each. Changes in the neutral, cortisone-mediated neurohormonal system set the stage for the overthrow of latency system-consciousness. The revolution in system-consciousness occurs when the level of sex hormones and growth hormone produce a change in the dominant hemisphere dopamine system that initiates the subjective regression that occurs in the egocentric stage of adolescence. According to Stevens (1979) the newly set hormonal modulation interacts with maturing amygdala circuitry, facilitating all forms of consummatory behavior. The maturation of the amygdala in puberty must also faciliate kindling. In the middle, narcissistic stage of adolescence, thyroid hormones begin to increase, which creates a feedback effect that increases the norepinephrine resources of the system-consciousness. Finally, at the end of the last, object-related stage of adolescence, we must assume that the neutral cortisone system re-establishes an equilibrium at a higher level of production. This feeds back to increase the serotonin resources enough to stabilize the system-consciousness for adult life.

The hypothalamus responds to the presence of male and female sex hormones as early as three months into the embryo stage. This response determines embryological sexual development as well as having a direct sensitizing effect on brain sensitivity to neurohormonal modulation throughout life. The sensitivity of the hypothalamus to circulating levels of sex hormones decreases during latency (Dorner, 1977). Conceivably, the newly matured cortex, which promotes the increased efficiency of operational thinking in latency, may be a major factor in the reduction of stress and consequently of adrenal-hypothalamic activation. I am willing to hypothesize that this low level of hypothalamic activation through the adrenal hypothalamic system decreases the hypothalamus' neurophysiological sensitivity to circulating sex hormones. That in turn begins to allow for the production of more sex hormones through a reduction of hypothalamic inhibition. At puberty the increased output of sex hormones facilitates energization of physical and psychological growth processes.

In prepuberty, increased levels of FSH and LH secreted by the hypothalamus in response to the newly heightened levels of sex hormones cause the genitals to mature at a faster rate. Kestenberg (1980) comments on the activation of the hormonal maturational process:

> In prepuberty, there is a significant decrease in hypothalamic sensitivity which continues until the final attainment of a new set point of the hypothalamic gonadostat. Through the mediation of gonadotropine-releasing factors in the hypothalamus, large quantities of steroids can be secreted before the feedback is set in operation through which a pituitary secretion of gonadotropine effects a lowering of gonadal hormonal output. In prepuberty, the irregular production of hormones seems to indicate that the hypothalamic gonadostat has lost its childhood sensitivity but has not attained a new equilibrium as yet. (p. 231)

I believe that the increased output of growth and sexual hormones triggers an extensive neurological development that begins in the dominant hemisphere.

During mid-adolescence testosterone production amplifies in males, developing its highest rate of increase. As we now know, the sex hormones act as direct neuromodulators, pushing the neurotransmitter system to higher functional levels with increased energization. Doering (1980) comments on the augmentation of testosterone in mid-adolescence:

> The most dramatic testosterone increases are seen in midpu-

> berty: 0.6 ng/ml to 2.2 ng/ml within three hours of onset of sleep. (p. 260)

The sexual interest and maturation of girls in mid-adolescence is also related to the increased level of sex hormones.

The overall effect of the increased level of circulating sex hormones and growth hormone in early and mid-adolescence appears to contribute heavily to the increased action drive pressure and the shift to dominant hemisphere activation. Carey and Diamond (1980) find, for instance, that at puberty the nondominant hemisphere perceptual function of facial recognition that is so highly developed in latency, actually decreases. Similarly, the ability to identify persons through voice recognition, a nondominant hemisphere function, also decreases.

The smooth, nondominant hemisphere regulated speech, movement, and expression of latency breaks down at puberty. The pubertal adolescent reveals uneven, awkward initiation of movement, speech, and facial expression (Kestenberg, 1980). Cognition also shows the shift of ego functions to the dominant hemisphere mode of organization as formal thought develops in early adolescence.

Cognition in Adolescence Adolescence transforms latency cognition, as Piaget points out, through the development of a new cognitive process in which a spectrum of hypothetical possibilities is superimposed on whatever happens to exist in reality at the moment. The development of what Piaget calls *formal thought,* in adolescence, recapitulates subjective egocentric process overwhelming earlier reality accommodative processing. Inhelder and Piaget (1958) summarize the cognitive changes that earlier in childhood moved from egocentric stage to reality-centered stage.

> Even at the sensorimotor level, the infant does not at first know how to separate the effects of his own actions from the qualities of external objects or persons . . . Later he differentiates his own ego and situates his body in a spatially and causally organized field composed of permanent objects and other persons similar to himself. This is the first decentering process; its result is the gradual coordination of sensori-motor behavior. (p. 342)

They describe the second egocentrism:

> . . . when symbolic functioning appears, language, representation,

> and communication with others expands this field to unheard of proportions and a new type of structure is required. For a second time egocentrism appears . . . this time the lack of differentiation is representational rather than sensori-motor. When the child reaches the stage of concrete operations (7–8 years), the decentering process has gone far enough for him to be able to structure relationships between classes, relations, and numbers objectively . . . the acquisition of social cooperation and the structuring of cognitive operations can be seen as two aspects of the developmental process. (p. 342)

Finally, the adolescent egocentrism transforms cognition once again, requiring a new decentering process:

> . . . when the cognitive field is again enlarged by the structuring of formal thought, a third form of egocentrism comes into view. This egocentrism is one of the most enduring features of adolescence; it persists until the new and later decentering which makes possible the true beginnings of adult work. (p. 343)

Formal thought is manifested in the classical, philosophical adolescent: does a sound in the forest unheard by cognizing ears really exist?

The first two stages of adolescence develop the "if-then" proposition. The function of agency is the first identity function to expand to conceptual level. "If" portrays agency in which hypothetical action goals are legion. The early adolescent is filled with grand inclinations. In the next stage of adolescence the cognitive functions of self and object-self are raised to conceptual level. This is the stage in which the adolescent discovers the truth of the proposition "I think, therefore I am." Just as his or her love shifted to the narcissistic object, so cognition shifts to the conceptual abstraction of one's being.

We know from Luria that the maturation of cognition requires a balance between action goals as hypotheses, and specific salients that must be present for the culmination of action-programs. In the final stage of adolescence the "then" portion of the problem-solving proposition reaches the stage of *social significance.* In adult life actions become significant only when they occur appropriately within a social context. This last stage of cognitive development requires the elevation of the object function to a social conceptual level of significance.

In his chapter on *Thinking Processes in Adolescence,* Keating (1980) emphasizes the relevance of all the meta-processes to cognitive maturation. He points out that all cognitive processes must be elevated to an objective meta-level of conceptual organization.

> The common thread is the individual's awareness and knowledge about cognitive activity itself and about the mechanisms that make it more or less efficient . . . Certainly the adolescent shows great metacognitive sophistication but also seems to expend considerable energy on internal cognitive regulation, occasionally to the point of elevating concern with the form of cognitive activity above the substance.
>
> . . . the second way in which this aspect of thinking shows a change in adolescence . . . is . . . increase in introspection . . . The third way . . . is more directly related to Piagetian theory. It involves . . . the performing of operations upon operations. (pp. 214–215)

To be mastered, the metaconceptual level must be synthesized and recognized.

Adolescent Emotional Development The egocentric substage, extending from approximately age eleven to fourteen, with an earlier onset in girls than in boys, centers emotional process on the fantasy that accompanies the first physiologically matured orgasmic masturbations. This experience, full of pleasure and pain, ends puberty as it kindles subjective transformation. The narcissistic substage of adolescence extends from fifteen to eighteen, with individual and sex-related variation. It centers on the experience of falling in love with another person who is a mirroring external object-self. At this time one's own subjective intensity (first experienced as a perfect inner possession) begins to seek an object that embodies and represents it. One thinks of Stephen Dedalus and the bird woman. At this stage, the object-self is experienced as manifest in another person. This identificatory narcissistic phase of the growth process might be called *primary hemispheric revision.* The narcissistic love object is cathected through the nondominant hemisphere. This experience represents a process of neurological fusion. By cathecting the nondominant hemisphere with newly matured action drive the individual begins to break away from subjectively centered narcissism, and begins a romance with the object world. The third and final substage of adolescent identity formation, the *object-related* period, extends from eighteen to twenty-one or beyond. This phase includes traumatic experiences of identity crisis around the assumption of a fully adult role.

Part of the lifelong experience of identity resolution through the working of biologically mandated fantasies, the *central masturbation fantasy* of adolescence externalizes sexual relationship from the family to the real world. The fantasy itself, an identity synthesizer, develops from egocentric involvement with genital sensations to a goal-oriented ambitious enterprise that takes the reality of the object into account.

As Lauffer (1976) writes:

> During adolescence, oedipal wishes are tested within the context of the person's having physically mature genitals, and a compromise solution (between what is wished for and what can be allowed) is found; it is this compromise solution which, within the variations of normality, define the person's sexual identity. *I therefore see the main developmental function of adolescence in the establishment of the final sexual organization*—an organization which, from the point of view of the body representation, must now include the physically mature genitals. (p. 298)

Lauffer goes on to define the "central masturbation fantasy" as a universal phenomenon that arises out of the oedipal period that contains the regressive pathways of gratification in masturbation and that structures the sexual identifications.

He points to the final organizing revision of this central regulating fantasy in late adolescence:

> If we examine the direction of the libido and the relationship to objects as expressed in the fantasies, we will find that the libido is object-related, even though the gratification is of a narcissistic or autoerotic nature; at the same time, the masturbation fantasies of adolescents, especially late adolescents, include the active seeking of a sexual love object. (p. 303)

Thus, the object-related quality of the central organizing fantasy of late adolescence alters the intrapsychic world of the late adolescent in the direction of a preparation for adult-form object-relations, at the same time that it regulates the flow of autoerotic and narcissistic aims.

Identity Crisis in Late Adolescence The end of adolescence restructures the central organizing fantasy. The restructuring restores the nondominant hemisphere to the forefront. As the late adolescent develops goals that relate to the object world and to the world of work, he structures these goals around a renovated version of the *family romance fantasy*. Since this fantasy structures identity integration and synthesis, it might more properly be called a construct.

Freud described the family romance fantasy as a mythologized life story in which the individual imagines himself as having a different set of parents. This fantasy operates to make cohabitation possible in adolescence by allowing the adolescent to enjoy the beloved's similarity to his real parents. At the entry to adult life, this aspect of personal mythology takes the form of an image of object-self as hero or heroine. The fantasizer imagines that he was born to redeem both himself and his parents, who were themselves undone by evil circumstances. The impulse to social reform originates here.

In neurological terms, this construct relates to nondominant hemisphere functioning and to the organization of the personality. The imaginative alternative element in this construct paves the way for a change in the nondominant hemisphere object-self and object representations. In the final stage of adolescence the adolescent must remodel the object-self to perceive oneself as an adult in the world, not as the child of one's parents. Freud (1909) emphasized this fantasy's value to the adolescent:

> The liberation of an individual, as he grows up, from the authority of his parents is one of the most necessary though one of the most painful results brought about by the course of his development. It is quite essential that the liberation should occur and it may be presumed that it has been to some extent achieved by everyone who has reached a normal state. Indeed the whole progress of society rests upon the opposition between successive generations. (S.E. Vol. IX, p. 237)

Freud says that a too avid devotion to earlier versions of the family romance promotes a failure to break completely from the parents. The fantasy must point the way to the adult alternative to remaining a child. In normal development, however, the family romance fantasy is an organizer for the instinctual life of the late adolescent who is on the threshold of adult life.

> . . . a quite peculiarly marked imaginative activity is one of the essential characteristics of neurotics and also of comparatively highly gifted people. This activity emerges first in children's play, and then, starting roughly from the period before puberty, takes over the topic of family relations. A characteristic example of this peculiar imaginative activity is to be seen in the familiar daydreaming which persists far beyond puberty. If these daydreams are carefully examined, they are found to serve as the fulfillment of wishes and as a correction of actual life. They have two principal aims, an erotic and an ambitious one—though an erotic aim is

> usually concealed behind the latter too. At about the period I have mentioned, then, the child's imagination becomes engaged in the task of getting free from the parents of whom he has a low opinion and of replacing them by others, who, as a rule, are of higher social standing. (p. 238)

The adolescent permutation of the oedipal family romance fantasy is therefore the myth of the romantic hero who overcomes his ordinary, real beginnings to change society in the direction of his own romantic aims. Mythological themes in culture—relevant stories, songs, movies, and the like—signify that the male adolescent's mythological hero is often the prime type who rescues the female adolescent from her humble origins. Similarly, the female fantasy often involves elevating a humble lover to the girl's own high station.

In late adolescence, the egocentric imaginative system (that contains all the autoerotic and narcissistic love wishes represented in the family romance construct) animates a quest for a real love object as well as for the means to change society. The identity crisis of late adolescent life begins as the mythology of the new object-self collides with the stark reality of the adult necessity of finding a means of economic survival. To achieve success in adult life, to work and to start a family, the late adolescent must learn the culture's language of social signifiers.

The final, secondary revising stage of adolescent identity transformation requires transformation of the superego. Signal codes of the real world must be structured in conceptual forms for processing social, cultural and economic information. The cognitive system must attain a full level of abstract conceptualization. The adolescent must grant society a conceptual reality that transcends both family romance longings and the family of origin structures. Society must be seen as autonomous. This granting of autonomy to the outstide world resembles the latency child's resolution of the oedipus complex.

VII: Homeostatic Systems of Adult Life (Age 21–70)

All of the hierarchical levels of functional synthesis we have discussed reach their optimal systematization in adult life. The adult system-consciousness is essentially a problem solving system. In this system's maturity the two hemispheres contribute equally to a fine problem solving synthesis that maintains intrapsychic and social stability in the individual and in society. Evidently social institutions owe their structure to an elaboration of various aspects of the adult system consciousness. The present section describes the system of homeostatic synthesis that maintains the adult system consciousness.

The adult method of problem solving is to equate the problem to be solved with the process of identity synthesis. Problems that can be solved by rote are solved in rote ways, but unique problems evoke feelings of identity, and these feelings are used to state the problem and to pose hypothetical pursuable solutions. The individual must divide difficult problems into parts that will fulfill a single goal. As Luria (1980) puts it:

> The person solving a problem must analyze its requirements, select the essential relationships, and discover the intermediate aims and operations by which these aims may be secured. Only by carrying through intermediate operations of this nature can he reach a final solution to the problem.
>
> The typical process of discussion or reasoning requires subordination of all the operations to the final goal, for otherwise they lose their purpose. It requires inhibition of all digressions from the final goal and, consequently, restriction of the whole process to a closed system, whose boundaries are determined by the conditions of the problem. (p. 580)

This goal orientation stands provisionally for all subjective identity strivings.

The two hemispheres tend to divide the work burden between them. When one hemisphere works harder, the other works less, or plays a supporting role. Similarly, the medial and lateral zones divide emphasis in functioning. When medial functions are going on, lateral functions tend to be diminished. Shimkunas discusses the "preattentive structuring" that determines the interplay and hemispheric balance. He believes the hemispheres may work in alteration, depending on how situational demands cause one or the other to be used. Shimkunas (1978) points out that electrophysiological studies have shown higher differential alpha blocking when the subject performs a task relevant to a particular hemisphere:

> The left hemisphere engages in alpha blocking (an index of attention) when subjects perform verbal tasks, and the right hemisphere blocks alpha rhythm when subjects perform spatial and musical tasks.
>
> Attention becomes a critical mechanism to hemispheric in-

> teraction. Hemispheres appear to collaborate by balancing attention between opposing directionalities in space and focus attention by preattentive structuring of the field, followed by selective focusing within the field in temporal sequences appropriate to the problem . . . In addition to this general progression, verbal-temporal vs. spatial-emotional characteristics of the specific situation would be operative in emphasizing which hemisphere is to be dominant. (p. 207)

In addition to the cognitive demands in problem solving, emotional demands also require reciprocal, interpersonal interactions to maintain adult system-consciousness stability.

Adult emotional relationships produce a kind of reciprocity in which the two individuals share stronger or more competent identity integrations. Thus, one aspect of the shared family relationship between husband and wife is the division of labor inherent in the sharing of identity functions.

The medial system of emotional integration is stronger and more developed in young women than it is in young men. This is not an absolute, but a relative statement about likely propensities for identity processing. For biological and social reasons women tend to center their usual sense of identity inside their body, their personal identity extending outward from their body to their mate and to their family of children. In interactional context, the woman's body comes to center the family, and the sense of family. The medial system of ego synthesis that provides intuitive knowledge of the world and of human relationships is more strongly developed in the woman. Consequently, in a complementary relationship the woman's role tends to acquire greater definition in the realm of medial functions and integrations. For biological reasons, the male tends to center his sense of identity in the lateral system of neuropsychological synthesis. This is to say that men emphasize those neuropsychological and ego processes natural to lateral synthesis, which they tend to emphasize in complementary relationships being mainly involved in propositional thinking and feature detection.

Social Homeostasis Many of society's legal and political, traditional and customary institutions help the individual and the family unit maintain the stability of their identity functions. To give one example, the institution of marriage (which has many other relations to identity functions that I will not discuss here) sanctifies the family romance fantasy. The ritual itself embodies fantasy and gives it social credence.

For the young woman, the marriage often revives family romance fantasies of becoming queen to her kingly mate and of taking title to the position which is ideally due her. The young man becomes the king or hero of his dreams and assumes the sexual privileges which have been his due all along if his *authentic*—in the family romance sense—child identity were known. Because the marriage ceremony allows the participants to socialize their oedipal and adolescent fantasies, it permits them to release their energies into their newly objectified identities, into the new family unit they comprise, and into society itself.

Within the family itself, the partners work out a new relationship that triggers another remodelling of identity. Kestenberg (1980) describes this process:

> By reproducing the spouse as a baby, the husband and wife . . . consolidate their relationship, which shifts from an intense love affair to a cooperative family enterprise. The form this cooperation takes is influenced by the example given by one's own parents and by current role assignments in the community . . . With each child, parents regress to the developmental level of their child, and as they do so, they help the child to progress to the next phase.
>
> In the average family, the mother regresses more, and holds and shelters while the father is more often the bearer of progression, a facilitator of letting go and separating. (pp. 76–77)

Hormonal Homeostasis of Adult Life The hormonal reciprocal modulation with the neural system contributes mightily to the high level stability of adult system-consciousness. The diurnal pattern of hormonal production is tuned to the problem-solving requisites of adult life. The relinquishment of problem-solving at the threshold of sleep as the adult swiftly regresses through the psychogenetic scale into the symbiotic entrance to sleep accompanied by the serotonin inhibition of higher cortical functioning releases the catecholamine systems from the energization of cortical functions. The uncoupling of the frontal system allows for a hierarchical regression to hypothalamic control of automatic regulatory processes.

The diurnal reciprocity of the drive neurotransmitter systems with the hormonal modulation systems produces the functional equilibrium of adult brain/mind process. Norepinephrine appears to regulate the production of ACTH, and presumably of β-endorphins and other peptides that share the same precursor molecule (POMC). The locus coeruleus, which integrates noradrenergic activity is a heavy neural source of control to the cells that produce ACTH Swanson son & Sawchenko, 1984). As the norepinephrine production falls from its midnight height, ACTH production begins to rise stimulating the high morning cortisol level. Thus cortisol and β-endorphin would be more available for problem solving and for survival oriented activity during the day when they are needed.

The somatostatin-growth hormone variation is also diurnal and tied into the adult problem solving cycle. Rose (1980) points out that GH is expended at those points in the day when action plan requirements are at their highest. Otherwise, the highest output of GH occurs during slow wave sleep. GH appears to respond to all three of the major neurotransmitter drive systems. Prolactin is another hormonal modulator that is in reciprocal relationship with the drive systems, and which operates on a diurnal cycle. I conjecture that it may act as a local messenger hormone signaling the hypothalamus that frontal system regulation is no longer present, for it is secreted soon after sleep onșet, even during naps. This conjecture is supported by Wagner and Weitzman (1980) who state:

> In addition to a tonic dopaminergic control of prolactin, there is conflicting evidence that serotonergic mechanisms may also be involved in the stimulation of prolactin secretion. Finally, there is evidence that prolactin "feeds back" on itself to inhibit its own secretion via an effect on the hypothalamus. (p. 238)

Thus the conjecture is warranted that prolactin is in a position to "inform" the hypothalamus when the drive activated system-consciousness is shut down.

Thyroid stimulating hormone and thyroid hormone are another dual feedback system that intermodulates the drive system in a diurnal control pattern. TSH is increased around the time of sleep induction, so that thyroid hormone is increased in its daytime availability. Like cortisone and growth hormone, thyroid hormone has an immediate effect that can be stimulatory to the energy enhancing cellular systems. Thus over the short term stressful conditions evoke a neuromodulatory response that enhances energy systems in the neuronal control of system-consciousness. Over the intermediate term thyroid hormone, like cortisone, increases the number and activity of beta adrenergic receptors. Thus periods of stress lasting for days produce an enhanced capacity in system-consciousness. The beta-adrenergic system may be considered the major energizing system, not only for neural tissue, but for many other tissue types in the body as well.

The sexual hormones also participate in the diurnal cycle of variation, and FSH and LHRH are hormones that in addition to neuromodulation and enhancement of energy functions of the neural cells they modulate, have general body effects of increased energization. Thus, we may conclude that in general the diurnal variation of system-consciousness work has an effect on the whole range of hormones that regulate body processes as well as neural processes.

Identity Transformation in Midlife The process of identity transformation in midlife from forty to fifty appears to be due to social as well as biological triggering effects. Possibly the most important and least cited reason for the identity crisis of midlife is the decrease in drive system neurons. The total energy availability for both action and reality drive decreases. For those persons who are primarily dominant hemisphere predisposed, the process of change will be a growth process; for those predisposed to nondominant hemisphere processing, the process of change will be primarily adaptational.

Stage I: Problem-Solving Failure During the late thirties and early forties it begins to become apparent that identity consolidation must set in. The adult is no longer in the position to pursue a diversity of career goals. The means of achieving success, and the diversity of problem solving modes also contract. Social pressure increases to define career goals once and for all. Persons who are not well placed in the job market have less opportunity to change their fate.

One major reason for the decreasing flexibility of problem solving may well be the decrease in the catecholamine neurotransmitter system's scope. According to McGeer, Eccles & McGeer (1978),

> . . . catecholamine cells in the CNS are the only ones where a definite and dramatic decrease of cell numbers with age has been shown . . . At birth there are about 400,000 dopaminergic cells in each substantia nigra. By age 60, the number has dropped to about 250,000 . . .
>
> Brody has found a similar cell loss with noradrenergic cells

> of the locus coeruleus. Here the numbers drop from about 19,000 for youths to about 10,000 for people in their 80's. The calculated lines of regression are significant in both cases, although the rate of loss in greater in the substantia nigra than in the locus ceruleus . . . (p. 404)

The loss of these cells begins at an earlier age than occurs in other neural systems and appears to be rather specific. There is no evidence for a similar decrease in the serotonin system. We can surmise, then, that in middle age while drive strength decreases, the ability to neutralize remains high, or even becomes proportionally higher than it was. If the decline in the production of sex hormones that takes place in this period (in women beginning in the menopause, and in men gathering force slowly in the later forties and fifties) sets the level of hypothalamic functioning at a lower level, then the need for highly active drive systems declines as the sex hormones decrease.

Stage II: The Regressive Phase of the Midlife Crisis As dopamine begins to decline, there may be at first a compensatory increase in action and sexual drive pressure, an attempt to reassert patterns of passionate subjective involvement. This may sometimes involve sexual acting out as part of a search for the foundations of subjective identity feelings. After a few years, sometime in the forties the norepinephrine system too begins to lose some of its potency.

During the midlife period some of the subjective goals that have been close to the sources of identity must be relinquished. At the same time, as the sense of energy available to process objective reality decreases, and signs of real physical wear in the body increase, a sense of vulnerability and mortality becomes continuous. The social world begins to seem less able to bind wounds and to hold unlimited promise of solution to problems. Disillusionment with the objective world brings the medial self into painful prominence. The effect of the increased congress with the narcissistic core of the ego is the re-emergence of the organizing fantasies and long term memories of earlier life stages.

The paradigm holds that diminished identity integration and diminished problem solving capacity lead to a traumatic limbic system response with a period of intense long term memory evocation. In this view, the recourse to memory in midlife is a result of massive hippocampal kindling due to failure in problem-solving on account of depleted norepinephrine and dopamine system resources. The solution to this identity crisis must consist of some reorganization of neurotransmitter resources so that existing or revised identity structures can more economically utilize the available resources.

Stages III and IV: Primary Process and Primary Revision Phases of the Midlife Crisis An influx of memorial data and the increased sense of mortality awareness characterizes this period, as writers on midlife so often observe. Cohler (1980) puts it this way:

> With the advent of middle age, time and memory are both reorganized. If the impact of the oedipal crisis was to turn attention away from the past, the midlife crisis has the opposite effect, and concern with the past becomes more apparent at this time than at any previous point in the past. (p. 172)

An historical account of one's own life becomes a necessary mythology in the formations of the revisions of middle life. Primary revision of identity occurs through a revived narration of family romance fantasy. Under the auspices of this fantasy the object-self is reconstructed so that it directs itself toward the fulfillment of middle age goals. Around the age of fifty, literary people often feel an urge to examine their pasts. This leads them to express an immortality fantasy that emerges in autobiography. This autobiography—whether it emerges in written form or materializes as intrapsychic reminiscence—lays the ground for the midlife revision of identity. For the change process to go on, new structures of identity must form. In the middle-aged person's case, the new structures of identity redistribute long term memory.

Cohler (1980) cites Vaillant and McArthur's (1972) report of middle aged men reinterviewed after 25 years:

> Now a group of highly successful business men and professionals, these men remember historical events in their own earlier lives in a way which is quite different from the way in which they remembered these events when interviewed during college. For example, men who reported remembering themselves as shy and isolated during adolescence, when subsequently interviewed as undergraduates, now at middle age report their adolescence in a quite different manner. (p. 165)

This does not, I think, represent retrospective falsification. Rather, the identity structure developed in middle age refocuses the earlier experience. The older men can see tendencies in the younger man which the younger man could only experience, being then unaware of the results of these tendencies and so unable to formulate the experience consciously until middle age.

The summary revision of memories of the object-self generalizes the person's potential to deal with life crises. The reorganization of memory and of personal identity allows the individual to get along with decreased neurotransmitter resources, literally with fewer dopamine and norepinephrine receptors. One's own youth is now retrospectively invested with courageous heroism. This idealization of one's own youth has the effect of consolidating diverse memorial structures.

The structure of the organizing fantasy is broadened in the midlife primary revision phase of the mind/brain process for change. Action tendencies that were only potentially goal-directed in youth are gathered under the aegis of more flexible and broader midlife goals. Goals are thus more conscious and more easily related to identity-strivings in middle age. Dopamine goes further when the conceptual categories of goal formation are generalized.

Stage V: The Secondary Revision Phase of the Midlife Crisis As the revision of identity proceeds toward completion, the nondominant hemisphere carries out secondary revision of system-consciousness objective structures of identity. In this light the middle-aged person develops increased interest in the means by which men of an earlier era managed to resolve their difficulties. The mature resolution of the identity crisis of midlife requires the creation of a more flexible sense of social structure, based on the belief that social structure is significant within an historical framework. Thus, the objective identity structures change in the same way that memories of youth were reconstructed, providing a flexible social framework that includes a sense of the remembered past as historical.

Cohler comes to conclusions that are typical of the resolution and reshaping of identity in this era:

> . . . there are a number of developmentally salient conflicts which are called forth at particular transition points across the life cycle and which have the effect of reorganizing the personal biography in ways which were not possible earlier in the life cycle. The task is to understand this process and the means by which a consistent personal history has been created as a result of these successive transformations. Increased awareness of this process is

> among the most effective means of fostering the capacity for self-observation and for increased neutralization. (p. 187)

VIII: Mind/Brain Economy in Old Age (Over 70)

Old age produces a marked shift to right hemisphere processing. The system-consciousness in old age must economize its functions to produce a well-balanced synthesis and integration of information under almost all conditions, for old age must avoid stress. The old person suffers particularly from the stress response that occurs when cognitive processing begins to fail. In youth a person can afford the shift to stress response with its attendant increase in growth hormone, cortisone, and thyroid hormone production, which increases the availability of neurotransmitters and increases the likelihood of successful problem solving. An old person cannot well encompass the metabolic reorientation, because the shift in resources that accompanies the increase in neurotransmitter availability is accompanied by a sacrifice of other metabolic processes.

The emergency fight or flight response is very impaired in old persons because of decreased epinephrine production in the adrenal medulla. In old age the stress response can lead to various chronic organic illnesses.

In order to minimize stress mechanisms the system-consciousness of old age has recourse to the highest level of generalization that can be structured in epigenesis. We have already seen that in the end stages of the psychotic syndromes individuals resort to premature mobilization of the cortex of generality, especially nondominant hemisphere processes of generality. In old age, the recourse to right frontal regulation is facilitated by the neuroanatomical fact that the right frontal hemisphere tends to be considerably more capacious than the left. To be sure, in terms of gross brain weight, the left hemisphere's temporal-parietal cortex volume more than makes up for this. As we know, the right hemisphere ultimately specializes in control, while the left hemisphere specializes in the production of meaningful language. To insure stability in old age, objective identity functions of control and subjective formulation of speech become pre-eminent. Thus, old age exaggerates the lifelong lateralization process, while it uses the capacious frontal portion of the right hemisphere to ensure feelings of familiarity in the environmental context. As old age processing shifts to the highest level of generalization, and to the more medial form of right hemisphere processing, the tendency to conservation in old age is maintained by a greater reliance on generalized self integration. This emotional affective channel becomes equated with the sacred in culture. The social function of faith in a supreme being tends to become equated with object-self identity in old age.

The shift to generalized right medial processing is the end stage of a developmental process. During adult life the right hemisphere increases in weight and capacity relative to the left. Bondareff (1980) describes the relatively greater impoverishment of the dopamine system in old age and the death and decline in acetylcholine system facilitating neurones that is particularly marked in the left superior temporal area and the left prefrontal area. Indeed, old persons show EEG evidence of abnormal slowing over the left temporal areas (Michalewski, 1980). The decline in sexual hormone production to very low levels sets the hypothalamus to a lower level of activation of consummatory processes. For all of the above reasons, the reliability of processing shifts to the right hemisphere.

The relative failure and decreased sensitivity of sensory end-organs, and the slower neurotransmission of sensory information all combine to produce a greater reliance on the tried and true features in perceptual composition. The old person institutionalizes action patterns. The reliance on hierarchical generalization—approximations based on life experience and wisdom—is a conservative system-consciousness measure that insures the least possible stress. All of the mechanisms for restoring system-consciousness capacity when it is used up, except for the cortisone-serotonin axis, are severely depleted. Growth hormone, thyroid, sex hormones, and epinephrine production are relatively depleted compared to cortisone production (Renner & Birren, 1980). When cortisone is secreted at higher levels, the aging process increases in tempo (Everitt & Huang, 1980).

Welford (1980) details the generalizing problem-solving strategy that older individuals tend to adopt.

> A realistic reaction to an incipient fall of capacity is to ensure that all detailed strategies of performance are as efficient as possible so as to maximize the demands that can be met by the capacity still available. One adjustment for this purpose . . . is to develop routines that enable large sequences of action to be carried out as unified wholes in which each item flows from the last with a minimum of fresh decision. One can also adjust by ordering sequences so as to cut out unnecessary actions . . . These efficiencies come as the result of experience and can be regarded as ways in which experience compensates for age changes of capacity. (p. 209)

Emotional Memory and Old Age The midlife reorganization of memories is extended in old age. In line with its hierarchical generalization of identity, the nondominant hemisphere medial channel triggers a final major adaptational change in system-consciousness that reorganizes the conditioned (or generic) memories of a lifetime, generalizing them, i.e., making conditioned experience apply ever more widely. Old age wisdom is a result of the generalization of a lifetime of memory and experience. The individual experiences the re-emergence of the layered organizing memories and basal fantasies of a lifetime. Schonfeld (1983) describes the benefits that accrue to the aged from their reliance on generalized generic memories:

> Generic memories themselves are formed from repeated series of episodic memories, and each episodic memory is a potential generic memory—hot stoves, burns, thunder follows lightning, shop B comes after shop A, red light means stop . . .
>
> Generic memories of concepts and environment are invaluable in coping with relatively stable physical surroundings, and their importance probably increases with age. They allow reliable predictions to be made in our ongoing activities. (p. 219)

On this basis I argue that the generalization of long term emotional memory in old age serves to preserve identity structure. This emotional generalization is the core process that determines the course of object relations in old age.

One is confronted in old age with a daily reduction of cognitive and physical abilities. This problem solving failure produces a necessity for changing both lateral and medial objective functions of identity. We may conceptualize this as the kindled work of mourning for the object-self qualities that have been lost in the course of life. Of necessity, this work of mourning evokes, as it did in King Lear, every bit of the fantasy life and of the memorial basis upon which the core of objective identity is built.

Trauma in Old Age The realization of aging and its consequences leads to an experience of a loss of the object-self. This profound loss kindles mourning that revives all the medial sources of memory upon which objective identity were built in the life process. At the beginning of this period of traumatic change the object-self is experienced as separate from identity.

Freud (1919) described in *The Uncanny* the object-self split off from the sense of synthesized identity that appeared in the prelude to his own old age:

> I was sitting alone in my wagon-lit compartment when a more than usually violent jolt of the train swung back the door of the adjoining washing-cabinet, and an elderly gentleman in a dressing gown and a travelling cap came in. I assumed that in leaving the washing cabinet, which lay between the two compartments, he had taken the wrong direction and come into my compartment by mistake. Jumping up with the intention of putting him right, I at once realized to my dismay that the intruder was nothing but my own reflection in the looking-glass on the open door. I can still recollect that I thoroughly disliked his appearance. (S.E. Vol. XVII p. 248)

Freud discussed this incident as an example of fearing one's "double". "Double" is indeed a good conceptualization of the traumatized object-self of old age. The incidental imagery of the door between the two compartments represents the object-self and the object, alternative components in the reality sphere. The object-self splinters and fragments in the entrance to old age. Doubling is seen in the old person's tendency to use the other as a prosthesis for the object-self. This form of mirroring transference appears to be common to old age.

The work of therapy in old age is to help resolve the fragmentation of object-self identity. Grunes (1980) describes the therapeutic work and the context for that work with the aged:

> . . . For the aged person who suffers a rupture in the sense of the historical self . . . the re-establishment of this sense of continuity is a primary therapeutic goal . . . Patient and therapist must work to recover past memories, and this work must be an active therapeutic maneuver involving interventions and reconstructions of such memories for the patient . . . The patient, bewildered and in need of touchstones, can find, with the uncovering and attempt to re-establish memories from his own life . . . a recathexis of his own past as a sufficient organization of the historical sense of self, perhaps less subtly organized but organized nonetheless to function as a unified personality. (p. 547)

This statement of therapeutic intent describes the intrapsychic work the old person must undertake naturally in remodelling identity in old age, in order to insure stability of system consciousness, for the breach in identity that is a product of traumatic splitting is repaired through a process of generalizing the historical imagos liberated by the splitting process. The re-emergence of past identity themes can be used for a generalized reconstruction of emotional identity themes, which need not be accurate, but which need organize the residual capacities. The broadening of outlook which we equate with *wisdom* comes from the necessity to generalize reflection on experiences to the point that all past experience continues to have utility.

Pollak (1980) gives similar testimony to the issues that must be resolved in the therapeutic analysis of the aged:

> . . . I have found the focus on the mourning-liberation process to be of great importance. The basic insight is that parts of self that once were, or that one hoped might be are no longer possible. With the working out of mourning for a changed self, lost others, unfulfilled hopes and aspirations, as well as feelings about other reality losses and changes, there is an increasing ability to face reality as it is and as it can be. (p. 576)

Secondary Revision in the Aging Process The family romance construct undergoes its final secondary revision in old age. There is a delicate equation of the self with God that achieves social support and emerges as a belief in personal immortality. Just as the narcissistic externalized object-self seems to encompass the identity aims of midadolescence, in old age God contains all of the nondominant hemisphere identity functions. The belief that man is the son of God is tantamount to an identity construction in which self, object-self and object functions are unified. The fantasy of going to heaven, which is culturally and religiously endorsed, embodies this ultimately conceptualized generalization of right hemisphere identity functions. The construct of immortality is a distillation of the possibilities inherent in the nondominant hemisphere portion of the system-consciousness. Faith and belief in God is the ultimately effective denial of objectified death. It cancels the stress of physical and mental decay as it protects against social alienation.

As a generalized metaphor for identity fused with reality, God is a distillation of the nondominant hemisphere system-consciousness at the superordinate hierarchical level. The resort to God as identity prosthesis is a resort to the construct that the mind itself, the system-consciousness, can remain synthesized and intact forever. Thus in old age, a belief in God, or one's immortality through progeny, or through works, are all basic biological-social constructs that regulate and preserve the constituents and bases of consciousness.

Chapter 6

PSYCHIATRIC PARADIGM TREATMENT PRINCIPLES

Pharmacotherapy and psychotherapy do not exclude each other. This is perhaps common knowledge, for we use it when we tell a patient that the stress he or she has undergone has been so prolonged and intense that it has exhausted neurotransmitter resources. This, we tell the patient, is the reason for giving him or her particular drugs. To be sure, in specific cases our explanations will be far more detailed and will include an explanation of the causal development of the patient's illness, so that the therapeutic alliance will benefit from an increased scope of mutual understanding. Thus do we combine, in practice, a therapeutic approach that includes the brain's resources and the mind's experience. If we understand the genetic layering of the mind/brain, and side neither with the mind nor with the brain in making our interventions, we may knowingly employ psychotherapy, pharmacotherapy, or both in combination. The choice must depend on the nature of the syndrome to be treated and on therapeutic goals. The present chapter offers a view of treatment possibilities that is based on the structural paradigm.

Epigenetic Principles

We have been developing an understanding of psychopathology and psychopathological syndromes as originating in distortions in the formation and epigenetic transformation of the system-consciousness. This approach has taken into account the fact that the intrapsychic system creates new consecutive layers of consciousness in each consecutive period. We have seen that biological conflict inevitably produces a process transformation. Identity integrations are constructed in the beginning of each life stage, and these are inevitably remodelled during the unfolding of each developmental period until the current integrations can no longer be successfully remodelled. This impasse comes about through the inherent biological tendency for change.

As the psychopathology in any life period will take form around the processes of stage formation and transformation, we must diagnose with this in mind. We do currently diagnose left hemisphere triggered psychosis as schizophrenia, and right hemisphere triggered psychosis as major depression, but no such consideration holds in the diagnosis of children. Undoubtedly some autistic syndromes are due to dominant hemisphere triggered malformations of subjective identity integration, while other so-called autistic syndromes must be triggered by right hemisphere depressive failures in internalization and identification processes. A similar epigenetic/hemispheric approach would facilitate diagnosis of old people for epigenetic factors are clearly manifest in the psychopathology of the aged.

Character Distortion in the Aged

The often noted increased frequency of "paranoid" thinking in old persons relates to the fixity and egocentricity of their personality structure. The old person is reality-bound, epigenetically programmed to involvement with the first two stages in the adaptive process, problem solving, and reframing with recourse to long term memory. Therefore, in situations of conflicting interpretations of reality or in confounding interpersonal relationships, the old person is disposed to see oneself as the arbiter of reality—one knows what is going on, and why. The amounts to a deformation of character in the direction of paranoid character structure.

Old people's depression may also be related to the need to maintain the reality-oriented problem solving mode. When psychic trauma occurs, and the old person has a tendency to withdraw, he or she uses massive denial to preserve the resources to reintegrate one's world, for the old person's norepinephrine sources are vulnerable through the depletion of old age. For the aged, depressive response is more conservative than an organismic one. If one uses too much energy in attempting to evaluate and redefine one's context in face of insoluble problems, exhaustion of the serotonin system can produce catastrophic indentity-system disintegration. Once the compensating system for organic brain damage has become disorganized, it may, like Humpty-Dumpty, not be able to be put back together again. For, if the higher level of synthesis of consciousness is lost in old age there is no refuge, no unplumbed organic resources of the cortex to be accessed. If this is the nature of the intrapsychic death to be feared, then a severe depression—even with rumination about the end of life and about emptiness—is to be valued beyond the loss of capacity to produce such a rumination.

The Process of Psychotic Dementia

Senile or presenile dementia caricatures the aging process. In the beginning of dementia, failure to include each step in the work process disrupts the dominant hemisphere's ordinary work. As nondominant structures that key identity functions begin to follow suit, long term memories flood the mind and take the suffering person (and those around him) by surprise. The process of identity disintegration can be quite speedy. Old identity themes—first from earlier adult life, and finally from childhood—organize the patient's enterprise and conversation. After this breakdown in upper level identity integration and synthesis, lower level ego functions begin to lose their own structural integration. The body begins to become an alien object; we may say that the object-self becomes alien. At the same time, action sequences deteriorate further. They are invaded by instinctual organizers. The patient is found picking at pieces of lint. Finally, in the last stages, the ego functions deteriorate into their neuropsychological components. Speech perseverates, with echolalia. Meaning and salience disappear. Emotional processes lose their cohesion. One sees affect displays of crying, and emotional displays of laughing.

A final massive defensive phenomenon stays the progress of the hierarchical organic and psychological disintegration.

> . . . these advanced Alzheimer patients . . . sit . . . in front of a mirror and talk . . . to themselves . . . When the demented person begins to display this and other bizarre behavior and is obviously becoming quite out of touch with reality, his dementia has reached the point of organic psychosis. (Strub & Black, 1981, p. 127)

The mirror sign is one last attempt to hold on to the reality function of identity by making the externalized object-self equivalent to all of reality. The object mirror image acts as the last semblance of reality that can integrate the nondominant hemisphere identity functions. This function stands alone against the disintegration of all identity integration at a system-consciousness level: the intrapsychic equivalent of death.

Neuropsychological Principles

In every diagnosis we must determine both the progressive nature of the syndrome and the hemisphere that initially fails in its problem solving function. Persistent unilateral hemispheric failure initiates the major syndrome. Inevitably the other hemisphere will be affected and produce part of the syndrome ontology. This effect will be overdetermined by identity changes occurring in the system-consciousness. Thus, to take schizophrenia as an example, after the initial breakdown in left hemisphere subjective identity functions, the right hemisphere will be more affected. Feeling of depersonalization and derealization will accompany destabilization of these objective identity structures. Persons with persistent failure in loving, narcissistic individuals, will similarly recruit a severe nondominant hemisphere response of intense guilt shutting down future expectations of loving. Depersonalization and derealization is more likely to occur in the twenties when the remodelling of character structure is still more active, while symptoms of guilty withdrawal are more likely to emerge in the thirties after a decade of love failures.

We must also diagnose the degree of neurological chronicity in the syndrome. As we have seen, as people approach their midthirties they tend to rely on high level identity integrations. The cortex processes social relations. Indeed, deep advancement into this cortical hierarchical development conserves system-consciousness energy during the aging process. In both schizophrenia and major affective illness (both premature aging processes that rely on syndromatic cortical generalization rather than a lifetime of appropriate social experience), as stages of chronicity supervene, the compensatory tendency emphasizes high level nondominant hemisphere identity functions that coordinate logical process. The over-generalized conceptual framework may have minimal relevance in unanalyzed social communication. It acts, nevertheless, as a vehicle for the registration of identity. The value of this process is that it is inclusive enough to relate the individual to society, since relying on upper level cortex, the individual passes into a realm of thinking that is more social than individual. Whichever side triggers the illness, the chronically psychotic individual welcomes the feeling that his or her identity has come to reside in some institution of society rather than in personal identity structure.

Tests of neurophysiological function will undoubtedly continue to increase in importance as measures of the degree of chronicity. It begins to become possible to develop a profile of neuromodulators that indicates the pattern of chronicity developing in each hemisphere. By developing a profile of the activation of ego functions the CEEG also should indicate the degree of chronicity. At least it should give an indication of the relative preponderance of time and energy spent in different zones of neural activity. Combined maps of brain/mind functioning should provide some automatic diagnostic indicators in the psychiatry of the future.

Psychoanalytic Principles

Fixation and *acting-out* are two related dominant hemisphere syndrome determinant mechanisms; and *trauma* and *repetition-compulsion* are two related nondominant hemisphere mechanisms that contribute to syndrome formation. *Transference* is a clinical phenomenon that manifests the effects of these four mechanisms. Thus far we have viewed fixation and trauma as two forms of life events that set up a kindling process that is instrumental in producing stage changes in system-consciousness organization. Now we must see how these processes operating *within* a stage contribute to distortion in further development.

We may define fixation as an unexpectedly deep form of satisfaction that is repeated in order to develop the action pattern that led to the satisfying experience. This mechanism obviously contributes to the formation of the perversions. Indeed, fixation is essentially related to sexual pressure. The young adolescent is particularly susceptible to sexual fixation, for in adolescence new action patterns are just being induced around hormonal developments and intensified feedback from the genitals. The early (three to six month) symbiotic period of infant development, when infant sex hormones are relatively high and the oedipal period when sex hormones also increase, are other likely periods for fixation. Theoretically, any period of intense psychological growth process with the development of new subjective means is a time of likely fixation. We may conceive of a *negative fixation* as an understimulation of modes of satisfaction that produces a poverty of subjective life.

From the neuropsychological point of view, fixation seems to be a schematization of dopamine use that becomes relatively enduring and that sets up a pattern of dopamine regulation outside of stage-appropriate identity function organization. Thus, we may consider the clinical concept of *subjective splitting* as the formation of a separate initiatory system—separate agency—functioning alongside of ordinary identity function. Such a separate initiatory development can account for acting out, as we can conceive of whole initiatory systems that escape the fate of ordinary system-consciousness development. Such a channel is also available for the exploration through regression of older motive states. We may take this concept of subjective splitting further, to the point of seeing separate and split off systems of initiation each with separate identity function. We may call this concept *subjective fragmentation.* When the center of initiatory identity does not hold, the person feels fragmented subjectivity, a dreadful state of narcissistic shattering.

Acting out therefore seems to be a tendency to use alternative subjective channels to revert to fixated modes of satisfaction developed in earlier periods of life. Acting out escapes the repression or defense that provides the barrier between stage developments. Not only sexual perversion, but other psychoneurotic and character disordered forms of satisfaction seeking may be enacted as a result of the fixated devotion to alternative forms of subjective identity. We may be aware that in the clinical enactment of the perversions, or acting out, the activity is carried out without conscious reflection and so circumvents the anxiety system.

The nondominant hemisphere has a similar mechanism for insuring access to past organization of identity. *Trauma* is an unexpected or ill-prepared experience of psychic pain that sets up a kindling process that forces the reception of new painful experience until the identity channels of the nondominant hemisphere have changed sufficiently to accommodate the new experience. Studies of stress-related trauma and post-traumatic stress reactions indicate that under these circumstances the perception of experience is fragmented by denial. The denial prevents access of the experience to ordinary identity processing. Thus a set of perceptual information is taken in outside of organized consciousness. Everyday clinical experience shows that painful experiences are segregated and tend to coalesce around split-off object-self and object identity fragments that are called "bad" representations. This produces a split in the objective functions of identity. If trauma is not too often repeated, or if it is not too intense too early in life, then these split-off identity structures may be gradually assimilated within the normal structures of identity. Repetition-compulsion comes into operation to assimilate experience that has not yet been integrated.

We may consider repetition-compulsion as both a tendency to repeat traumatic experience in milder forms and as a tendency to include previously denied aspects of the traumatic experience. Mastery of the denied and traumatic experience requires that it become subject to the organizing influence of ordinary objective identity, and therefore also accessible to recall. Clinically, repetition-compulsion may be seen in the re-experiencing of constellations of present day experience in past emotional terms, and in dreams as actual repetitions of traumatic scenes from the past. We may conceive of *negative trauma* as a life condition of too little novelty, leading to consequent maladaptive blandness in the personality structure. Both fixation and trauma, and the attendant mechanisms of acting out and repetition compulsion bring the past into the present.

The concept of *transference* may be used to understand the return of emotional configurations from the past both as they are transferred onto the present and in the special case of their being re-experienced in therapy. In therapy, the patient uses the therapist both as a mirror of subjective motives that arose in the past and as a dramatis persona derived from past emotional scenes of trauma. Depending on one's hemispheric predilections, a patient will tend to set the growth or the adaptive processes into motion when in a psychotherapy setting. The process of psychic change naturally comes into play, for psychotherapy exists to aid voluntary change of identity structure. Thus it must operate through the processes for psychic change. Considered as a whole mode of growth and adaptational elements, this process has a natural course which is encouraged by interpretation. If the patient is a nondominant hemisphere type of person who has suffered from trauma and who shows repetition compulsion, then the therapy will tend to produce profound identification process with the therapist in an attempt to remodel the objective aspects of personality and to overcome the effects of the past in this way. If the patient is a dominant hemisphere type, who has a great deal of fixation operating, and a tendency to act out, then the therapy will tend to follow the growth process mode of psychic change. In this case the patient will use interpretation to encourage regression in the service of uncovering the past fixating experiences, and the therapy will become an arena for re-experiencing the past fixating experiences.

Psychotherapeutic Experiences

We can conceive of *supportive therapy* as an attempt to help the patient stabilize identity functions that have become unstable through failure in problem solving. The method of doing this runs the gamut of social intervention, uncovering denied or repressed information, timely medication, and simple discussions of the common problems of adult life. We may conceive of insight-oriented psychotherapy as a set of methods for facilitating intrapsychic identity change.

The diagnostic consideration of which hemisphere is more involved in the present production of the symptomatology can be a guide to intervention. Let us consider the dominant hemisphere first. Many persons have a poverty of stimulation rather than a history of fixating experiences. Patients with such negative fixation feel that they have been left to develop on their own, with no particular limitations on their action patterns. They feel that they have had to discover their own methods of growing up. The parents in such situations will say, if they say anything, "You got yourself into this, now it's up to you to get yourself out." Such patients have been deprived of their opportunity for narcissistic maturation. In their therapy they need to have precisely this mirrored back in an understanding way that also describes what was missing. These patients will inevitably experience the therapist's understanding as a poverty of care, and the therapists mirroring as a re-experience of parental neglect. Therefore these patients will develop a symbiotic transference that leads them to believe that only the therapist can conduct therapy and have goals. The analysis of this transference will allow the patient to feel that his or her own efforts have indeed accomplished the work of the analysis as one's own efforts define one's own particular mode and means of loving and working. Finally, such patients must understand that everyone does need some mirroring and understanding. This is therapeutic because it reveals that the patient's needs and wishes are not unusually intense demands as they have been thought in the past.

One major principle for nondominant hemisphere targeted psychotherapy involves an attempt to help the patient reconstruct the fragmented sense of participation in reality. The tendency to avoid painful reality leads to a great deal of denial and disconnection in the patient's involvements. This must be understood, and hence explained, along the lines that maladaptive stress response, with denial of reality increases, rather than decreases, psychic pain. Then it is often necessary to help the patient reconstruct the attachments to reality. Many patients have become so full of denial (which may or may not reach psychosis-inducing proportions) that they have literally forgotten the pattern of attachments to family, friends, and society that has provided them with salience in the past. They may have gone off onto a life pathway that does not even feel like their own. This must be interpreted in the attempt to bring the patient back into the mainstream of his or her life. This line of psychotherapy encourages massive identification with the therapist as an expert on life. This form of identification must be understood by the therapist as encouragaing the patient's reconstruction of object-self identity function, permitting the formation of conceptual clarity about one's own life. Finally, the patient must come to know that he or she has redeveloped the personal resources with which to assess both one's own life and the social and political structure in which one finds oneself.

Pharmacotherapy Principles

Pharmacotherapy applies mainly to individuals who have developed a psychosis, a major depression, or a syndrome that is chronic and progressive. We must be wary of prescribing medication for individuals who suffer primarily from milieu-triggered exacerbation of their character disorders or minor anxiety problems, for they will tend to experience the tangible help they receive from drugs as fixating. The medication always induces some secondary gain that makes the drugs themselves and the experience surrounding their prescription into an artificial identity function buttress.

In all cases of acute psychosis our first concern must be to stop the progression of the illness before identity functions have been compromised. Our primary concern must therefore be that the patient be able to get one night's sleep, for staying awake (as happens in all psychoses), exhausts the serotonin system. The consequent lack of capacity to neutralize produces a regressive crescendo of hierarchical destabilization that represents a movement in the development of the syndrome. Sleep can be disrupted by dysfunction in either hemisphere. Both dominant hemisphere triggered *panic* (about the disintegration of subjectivity) and nondominant triggered *terror* (as experiences of unrelenting novelty and utter frustration) lead to insomnia. To check the sleeplessness, the physician may prescribe the appropriate gabaminergic reinforcing drug—one of the short acting diazepine compounds, for instance, in a dose sufficient to produce sleep.

The second issue is to deal with catecholamine depletion or destabilization. Hemispheric diagnosis determines which hemisphere has been depleted. A deep shame syndrome like schizophrenia and schizophreniform psychosis tends to deplete dopamine. Neuroleptic agents are therefore indicated. A guilt syndrome such as we see in the affective illnesses probably depletes norepinephrine and requires an agent such as a tricyclic antidepressant that first restores and then stabilizes the norepinephrine system.

Any psychosis will deplete the serotonin system, and at some point an attempt must be made to restore it. Lithium and some of the tricyclics or other newer antidepressant agents, and even some of the benzodiazepine compounds may produce this restoration. Some drugs, such as the still experimentally used piracetam appear to strengthen the system-consciousness. These drugs, the nootropics, may in the future, I surmise, stabilize system-consciousness and tend to prevent the outbreak of new psychosis.

We have become aware that any chronically stressful condition (including physical injury) can deplete the norepinephrine system. Syndromes secondary to injury, especially those with chronic pain, may benefit from the use of tricyclics. Organic assaults to the brain can also interdict the neurotransmitter systems directly, or upset the balanced regulation of the various systems. Therefore neuroactive drugs should be considered for the treatment of organic conditions such as stroke.

We must be careful, though, as the prolonged use of neuroleptics and tricyclics appears eventually to inhibit rather than to facilitate transmission of the catecholamine systems. This may contribute to chronicity by restricting the availability of these systems during periods of attempted repair of system-consciousness damage. Some of the recrudescent symptomatology in chronic schizophrenia, for instance, may be due to restitutional attempts to remake and remodel psychic structure and identity that has been damaged. The failure to have enough neurotransmitter available, or the reinstitution of psychopharmacology to block the recrudescence of symptomatology during periods of repair may prevent the person with psychosis from rebuilding psychic structure, thereby promoting chronicity. It is not unusual to hear a person talking about finally coming to know that he knows himself in such a vaguely abstracted way that we mistake the statement for more psychotic grandiosity and superiority, rather than hearing the side that announces successful reachievement of identity integration repair. The fact that such acute schizophrenic exacerbating drugs as amphetamine may alleviate so-called negative symptoms of schizophrenia in subacute or chronic stages, tells us that the lack of available catecholamines comes to play a role in chronicity. Thus, giving neuroleptics when a patient is actually trying to rebuild hierarchical identity function and structure only serves to force the patient into chronicity. The psychopharmacologist must therefore learn a more discriminating appreciation of phenomenology. For every syndrome utterance contains both a progressive statement of the intent to rebuild psychic structure, as well as a regressive expression of psychic disintegration. The therapist must hear both sides before settling on a course of medication.

Clinical Material

The case material presented here is not meant to exemplify complete treatment process. It is, however, meant to show that the structural paradigm can guide new approaches to the psychotherapy and psychopharmacology of the patients suffering from psychiatric disorders. I have therefore attempted to show how epigenetic principles and neuropsychological principles enter into treatment considerations. I have also tried to show how our ordinary guiding psychotherapeutic, psychoanalytic, and psychopharmacological principles can be enhanced through the use of the paradigm. The following case reports represent a selection according to the various diagnostic categories that we ordinarily treat in our clinical practice. Enough facts have been altered, though I hope in dynamically consistent, and diagnostically insignificant ways, to protect the identity and life history of individual patients.

Nondominant Hemisphere Triggered Psychosis

A Case of Mania Many patients have suffered incredible losses of family, jobs, health, and finally in taking prolonged flight, have lost connection with the roots of their objective structures of identity and have ended up in the hospital. One such patient, Ruth, was brought to the hospital by the police in a manic state, claiming that she and her "husband" were making a sex movie, and that she had been temporarily separated from him. Embedded in a peculiar bombast was her claim that she was also a stunt driver. Ruth was haughty, superior, and sarcastic, which is to say that she took great liberty in insulting all who attempted to talk with her. Like other manics, she picked out superficial characteristics of her victims, using these characteristics to begin a prolonged attack. She seemed to feel that any racial or ethnic slur was warranted by her, an attitude that inflamed the primarily black staff, as she delivered her oratory in a southern drawl. At the same time she exhibited the pathos of a totally lost little girl. The pharmacotherapy consisted of the administration first, of a neuroleptic, and then, after a week, starting her on lithium. The combined psychotherapy and social work consisted of an attempt to take every element of her story that might have been true, and to investigate the nucleus of reality in it, both intrapsychically and in a detectivelike fashion in regard to her real world attachments before coming into the hospital. For, like many chronic manic patients, Ruth had become a homeless wanderer living a fabulous inauthentic existence. Treating such patients within the usual guidelines of acute care time limitations can be all but impossible.

The stunt driving element led to the name of a California town where she had last resided for a few months before entering a mental hospital there. Calling the mental hospital led, through the recollection of a doctor, to her real name, which was not Ruth. Her boasts of being the star of a sex movie led to another person who lived in a southern city. Calling this man, and being able to refer to the patient by name, allowed us to reconstruct that that city was the first stop in her 2-year peregrination since leaving her three small children.

The patient had lived in a trailer in a small southern town with her husband and their three children. Her husband had simply abandoned her after beating her. In the course of his investigation the sheriff had raped her. In a panic, the patient gave her children, as if temporarily, to her husband's family and fled penniless to the city where, probably still caught in the acute stages of trauma, she did make the sex film. We may note that manic persons often become involved in sexual and financial escapades that are continuously adding fresh trauma to their state. After the movie people had used her, she went to the California movie capital where she briefly engaged as a stunt person. She was hospitalized as her decompensation became apparent, with an "atypical psychosis."

After drifting away from her aftercare in California she met a man who told her that he had contacts to make a sex movie in New York. Once arriving in New York, the man seemed to disappear.

In the 2-year interval since she had left home, traumatized and manic, the patient's object-self identity had become increasingly fragmented. Her denial operated to carry her in further flight from one identity refuge state to another. By the time she was admitted, she referred to a man she met briefly after her desertion by the man who had promised her a sex movie, as her "husband."

It is not unusual to find an unknown patient in a psychiatric ward. We might even disregard this fact in our zeal to arrive at a DSM III diagnosis, and, as I have noted, time is so short and resources so limited that we often cannot undertake the kind of investigation necessary to reconstruct the person's intrapsychic and social past. In such cases it is probably a frequent occurrence to treat the person as a schizophrenic. In the case we have been discussing, for instance, the patient would not have had a chance to recover without some therapeutic intervention oriented toward elucidating just how her denial had cut her off from her past. Even after the lithium treatment had been instituted the patient was hardly cooperative in the attempts at reconstruction. However, there does come a point in such cases when the therapist, by reconstructing a sufficiently coherent story of the life history, can induce a reintegration of the patient's objective identity functions. This is the point when recovery can begin. Lithium alone cannot promote this reintegration.

Indeed, the chronically manic patient with regressed and fragmented identity functions may require several months of inpatient treatment before all the links to the past life of the personality can be re-established. Social factors are therefore inevitably determining factors in the prognosis and outcome of such cases. The disposition in this case was possibly happy.

The patient agreed to seek treatment in a mental hospital in her own small southern town after a 2-month stay in the New York hospital. The standard treatment in this case would have been ineffective, as in cases of mania it often is. For the manic cannot form a therapeutic alliance in the short time during which lithium or some other serotonin restoring agent comes into operation. The manic must first have the intrapsychic experience of coming to know that he or she can rely on one's own object-self function before other individuals attain a reliable objective reality.

Cases of chronic mania with what appear to be delusions, but which are actually haughty generalizations of inflated object-self identity rationalizing extreme forms of denial, are not rare in my experience. As compared with the delusional structure of acute schizophrenia, or even of subacute paranoid schizophrenia, the structure of false ideas in chronic mania is looser, and the false ideas are evoked more casually, with the formation of new false ideas apparent in the ongoing communication to support superiority and denial. In fact, the habit of denial is seen in every piece of the flight of ideas that accentuates the object-self at the expense of the object. The delusional ideas do not maintain grandiosity (inflated subjective esteem) as they do in paranoia or schizophrenia, but superiority (inflated self-esteem). The fragmenting effect of the manic denial is constantly at work within the delusional ideas.

A Case of Endogenous Depression The patient is a sixty year old woman who developed her first formal outbreak of psychotic depression when she was forty. At that time she had finally, with the aid of psychotherapy, conquered her ambivalence about having children, only to find that she needed a hysterectomy. After the hysterectomy, in order to stave off feelings of depression and loss, she and her husband went to a foreign country, her husband's country of origin, to find and adopt a child. During this time they stopped having intercourse, agreeing to allege the cause as the husband's impotence. The impotence appeared to figure in their decision to choose a very lively one and one-half year old boy in the orphanage rather than a more sedate little girl who also captivated them. The child's age corresponded to the period of time that had elapsed after the hysterectomy. After 2 more years it became apparent that the child's lively qualities were due to hyperactivity. Indeed the child suffered from brain damage with petit mal epilepsy.

At the point when the child began school and was found to be somewhat retarded, the patient developed a strong death wish toward the child. At the same time she felt that she herself was damaged and that the child was an extension of her damage. When she began to have fantasies of killing the child, she felt suicidal. On one occasion she spanked the child and felt herself losing control. After that she developed an endogenous depression marked by extreme guilt and agitated depression.

The patient's object relations had been characterized since her childhood by a feeling that everyone had to love her. She was the youngest of eight children conceived by a mother who was tired of children and who did not welcome her. Rather, she was indulged at times and neglected at others. The family entered a period of difficult financial problems as she was growing up because her father developed severe diabetes, became semi-blind, and had to have his legs amputated. He could no longer work as a baker. The father, in his forced retirement, indulged the nine year old girl, but was not really interested in her.

As the father's decline and the mother's suffering bereavement increased throughout the girl's latency, first one older brother and then another began to treat the girl as an incidental sexual object. She felt gratified and special to be stimulated genitally by her brothers and to be enticed to play with their genitals. However, her secret guilt became an explanation to her of her mother's neglect. She also felt that her mother was a hypocrite who fussed over her sons while neglecting her. In her mind, she debased her brothers, feeling that their sexual activity with her made them really undeserving of the mother's affection. She believed that one day her mother and father would understand that she was really the best of the children, and she determined never to leave home until she had made her point.

She married, however, in her late twenties, after her father had died and mother showed little interest in her remaining at home. She transferred to her husband all her hopes of being valued and finding compensation for being undervalued. He was a university professor, but he had only an ordinary career, and as this began to become apparent, she shifted her wishes to having a child whom she would make into a doctor, as one of her brothers had been made by their mother. As her mother had not endowed her with a valued object-self, no more than the mother had of her own, the patient set about cathecting her son as a valuable externalized object-self.

We can see in this life pattern that the patient had vested her life preoccupation in nondominant hemisphere concerns. Someone else would always have to mediate reality for her and so make her special.

Because this never happened, every single object relationship took on the quality of early overidealization followed by disappointment, disillusionment, degradation, and resentment of the object. Nevertheless, she could never relinquish an attachment once one was made.

During the course of the years of treatment that followed her initial endogenous depression, it became possible to discern some rather typical characteristics in the onset of her acute illness. They began with strongly affective, ambivalent responses of anger, and fantasies of the objects of her anger dying. She "disposed" of her husband in this way as she was becoming ill. Typically, for a person with nondominant hemisphere ego function difficulties, she equated her husband's extreme nearsightedness with her father's diabetic retinopathy and concluded that he was developing the same affliction as her father. Then she wished he would die rather than become useless to her, as her father had become. After showing her anger toward him, she feared that she would precipitate his suffering a heart attack. He, for his part, was hypochondriacal, and amplified her fear. Thus, during the acute endogenous depression she also developed agoraphobic terror connected with leaving her husband and her son, for fear that they would be killed. During this period she developed early morning waking, rousing with a start. She often awoke out of a dream of the past, which it took her quite a while to shake. She suffered daily agoraphobic terror at the thought that she would be alone in the world with her damaged son. No plan for the day could provide her with the hope for gratification.

In the therapy sessions conducted as she was recovering from the depression the patient talked about the past as if it were still present. Her affective responses to her parents and her brothers and sisters were so fresh as not to have been diminished by the passage of time. On the psychiatric ward she clearly took the position of a very special child around whom the world should revolve. Interpretation of this wishful re-experiencing of her past in the present evoked an image of herself at the center of her family dancing in a circle around her, while she, the special young child, was the celebrated center of the circle. This motif, taken from a Jewish folk dance she remembered, was also an apt image for the reconstruction of her object world as her reality functions became reintegrated. With the help of tricyclic medication and with an interpretation of her transference to the therapist to the effect that he embodied all of her disappointments and disillusionments in the past with every member of her family, and that she was imposing her past on her understanding of the present situation, she was able to recover from her endogenous depression. Clearly, her therapist had become the central reality-mediating object who was at the same time acting in lieu of the intrapsychic object-self identity function allowing for a distinction to be made between the past and the present.

It is pertinent to describe her personality characteristics during a period when she was no longer suffering from ostensible depression. She had a wonderful ability to relate to the world as a "mensch." Her imitative qualities were very good, and she was able to identify with the characteristics of other persons very easily. She could render the world of the past in great detail. The identifications with her parents' Jewishness was very fresh. Every day, it seems, she would "hear" the Yiddish radio music her father listened to during her latency. The music made a motif for days, running and rerunning through her mind. She would find herself humming the tunes of the past to herself. Her friends valued her ability to entertain them with stories of the past, the old Jewish Baltimore they had shared. Everyone said she could be a writer, but she could not carry through any projects of her own. Instead, her life energy was concentrated on making herself valuable to others.

Even during her depression-free periods there was a loose structure of objective identity functions that indicates her readiness for another decompensation, given the intensity of stress to trigger such a response. Thus she showed masochistic and obsessive concern for her body image. She would pick at her skin, trying to perfect her body, at the same time deriving a really perverse masochistic pleasure in the feeling of spoiling her appearance. For her, squeezing pus or sebum was so tangible and drive-actuated that she easily equated it with the anal sphincter closure on the mass of her feces. In fact, she would withhold her feces for days, filling herself with a masochistic sense of inward badness, which was then delivered with much pleasure like the squeezing of a pustule. She took equal pleasure in squeezing a pimple on her husband's back.

The psychodynamic nuances of these body preoccupations were not lost on her. In fact she showed a fluidity of primary process thinking that allowed her to see the baby-feces-penis equation in concrete terms, easily evident in her dreams and in her associations. She understood correctly that she felt that her mother would have loved her is she were a boy, and that this was reason for her overt fantasy that her fecal column, for instance, could turn into a penis. She delighted to give her boy, as he was growing up, views of her own self-ugly body. She felt that as a woman she was ugly, but that she had a beauty in the soft fatness she developed that made her into a special child enveloped in her mother's flesh.

Her self-understanding was effective and curative, at the same time that it carried repetition-compulsion forward as a punitive mechanism. She was forced to recollect and relive the pain of the past. The work of the therapy centered on helping her to engage the present, as best she could, to be an effective mother to her child, to realize that she valued her husband as her close companion, and to help her to work appropriately in her job for money rather than for love.

Therapy with this patient has been undertaken on a chronic basis. No attempt is made to draw the therapy to a close after 13 years. On previous occasions such attempts have tended to precipitate further depression. The patient is also on a low dose maintenance tricyclic medication, despite mild recurrent extrasystoles that have required medication. Her husband has survived into old age without tragedy, and she feels very fond of him. He is no longer the image of her father. Her son has grown to be an adult. He gets job after job on the basis of his personality that mimics the middle class characteristics of his parents. He is, however, unable to do the work required. Even this does not throw the patient into depression. She is financially secure, as she has continued to work for the past 10 years while her husband is on pension.

The paradigm helps in this case by making an overall context in which the predominantly nondominant hemisphere symptoms can be understood. The patient's object relations are all ambivalent. This is necessarily so because every person represents an extension of her own object-self. Every person is judged by what he or she can do for the patient in her reality. Every person becomes a carbon copy of her mother who failed to pay her enough attention. Thus other people need to be reformed, to be made into good parents. We can see the patient even feeling that other person's bodies are extensions of her own self-image. Thus, she goes on a diet with her husband. When he fails, she has an excuse to fail also. She feels that her body is ugly like her father's after he was damaged, and like her mother's after she grew old. Those ambivalent body identifications explain to her the failure to receive enough satisfaction.

Chronically involved with the past, this patient's memory is a memory of deprivations and of vows for reparation. In remembering her mother's sacrifices and identifying with them, every image of her mother becomes a compensating treasure. Her mother's clothes and household objects become links to the past that are filled with reparation. Every person she meets is experienced as a potential treasure trove, who will die bequeathing her a fortune. The past as a deprivation filled with counteracting treasures is affectively alive in her own speech, as she repeats the Yiddish sayings of her mother in full prosody of intonation and affective reverberation. Thus her nondominant hemisphere social language constantly brings her parents near her.

Her long term memory and social language are nondominant hemisphere forms of neuropsychological processing. These images and representations of past reality have never been internalized and remodeled as true object-self.

Inasmuch as the task of the middle-aged person, particularly the person preparing to enter old age, is the remodelling of object-self, her heavy involvement in the past and in her deprivation traumas is used therapeutically to make an ever broader generalization of her identity. However, she never seems to come to the end of fresh generic memories, each one itself sufficient to warrant that she has been cheated and is in need of reparation. The emphasis of the treatment is therefore on the present. It is a relief to the patient to know that she attempts to hold on to her mother by treating her son as she wished her mother had treated her. Every interpretation or reconstruction ends with the fact that she is living now, and must make the most of her life in the present, that the rest is illusion. Thus the therapist becomes the spokesperson for the reality principle in her middle age.

This is the kind of patient who convinces us that there is a serious breach in her sense of identity reaching back to the roots of her mental integration. She maintains the potential for psychotic and endogenous depression through this breach. Her incompletely assimilated identification produces a fragmented objective identity structure, capable of bringing the past into the present as an interference with the integration of present identity. I believe that coming to the conclusion that this is potentially interminable therapy gives the therapist the correct position to take with the patient, namely, "where there is life there is hope," and, things have never gone so well for her in the past as they are in the present.

A Case of Dominant Hemisphere Triggered Psychosis A chronically schizophrenic man, now fifty, began treatment at the age of forty, never having had intercourse with a woman. He had, however, had a few brief homosexual affairs. At the beginning of treatment, in the first encounter, the patient declared that he could see no reason for telling me about himself, or expressing his thoughts, because he knew that I could read his mind. It soon became clear that he maintained the primary delusion that his body was not a man's body at all, but a woman's body, with a soft round belly that made him feel very reassured. He had few human contacts, although he did manage a low-level civil service job. He was referred after meeting a psychiatrist one day while walking his dog.

The patient had lived for some years in a single room apartment with his dog. He felt that his dog was ferocious and protected him from people. The basic interpretation was that because he could not bear to part with her, the patient felt that his body was his mother's. This led to an analysis of his symbiotic feelings of attachment to his mother. He felt humiliated when he thought of his mother's failure to relate to him as a person. The separation anxiety triggered rage reactions. The rage was attenuated by the fantasy that they shared the same body. During and immediately after the rages he felt that he could kill his mother. It became clear that to this man all women were figures of maternal symbiosis.

Another fixation of his action drive involved his taking the dominant position with a man. During his masturbation, he fantasied becoming part of a perfectly strong man's body. He sheltered himself in this illusion as he penetrated his male lover. Sometimes, during penetration, he felt his lover's body turn into a baby's. He equated this body with his own child body, which he had renounced by dominating his own baby brother. Thus, literally terrified of being hurt by men, and by women, in his perversion he extended the delusion of being sheltered inside his mother's body to sheltering himself in his lover's body.

Despite his schizophrenia, or alongside it, the patient courageously refused to relinquish his subjective identity and his wish for personal growth. He was able, with help, to articulate the feeling of so-called "atypical depression" that is often present in the movement from subacute to chronic schizophrenia, the feeling that his life was being wasted. The therapy went forward on this basis for a therapeutic alliance. During his first date with a woman, the patient had an experience of deep humiliation. The experience reminded him that he felt the same humiliation when he felt that his mother did not see him as a potential growing and developing person, only as a baby. This fixating maternal attitude is often found in schizophrenics' background. In response to his feeling of humiliation when his date questioned his sexual experience, the patient felt an increasing panic. Then in a rage, he shoved the woman away. She fell.

The therapy focused on his wish to kill this woman who had humiliated him, as well as his intense shame after shoving her down. He had to decide against repeating this potentially murderous impulse before he felt the subjective integrity to go forward in his development. His decision, a decision arrived at in his therapy, not to kill her, engendered increased feelings of subjective control and appeared to set off a period of intense psychological development. It seems that this conscious decision not to kill the object of symbiosis under any circumstances is the only way to ensure functional closure of the regressive channel back to the symbiotic fixation. For the first time, the patient's narcissism became organized, and he longed for my admiration. The mirroring understanding he received in the treatment gratified him enormously, and he felt that I became his external object-self.

At this point the patient joined a therapy group. There he met a woman who had once had a brief period of psychosis. Eventually they married. The patient worked consciously hard to satisfy this woman. At first when he had intercourse his mind was full of strong, able-bodied men. Later, his wife took on a crystal vulnerablity and in his mind her emotional state and body's state required masterful manipulation to satisfy.

Much of the weekly therapy over the following years related to narcissistic development in the Kohutian sense of the term. The patient manifested a paranoid sensitivity to acceptance and humiliation by other men at work. The cause of this sensitivity was not far to seek. He had never been wanted by his father, a narcissistic and self-preoccupied man who had taken the involvement between the patient and his mother as a permanent affront. He had, moreover, a brother 3 years younger who had gone his own way in life, abrogating family attachments. This left the patient with a feeling of humiliating abandonment by men.

The patient always suffered at coffee-break time because he felt no one wanted his company. One day he was invited to join co-workers. He soon felt, however, that people were making mocking allusions to him. When I questioned the reality basis for the patient's feeling of humiliation he went into a sustained rage. My questioning had broken the idealizing mirroring transference that the patient was using as a nucleus of identity development. He left the therapy for 6 months. When he returned he convinced me that his so-called paranoid attitude was founded in reality, that his sensitivity to humiliation was correct, and that he was right to withdraw from a large part of the social framework in his office.

He felt justified in concluding that the whole social group did not take their work seriously. Subsequently he engaged himself in his work, seriously, and for the first time, narcissistically. He realized, with some interpretive help, that a subjectively selfish motive, such as wanting to get ahead or to enjoy one's own abilities and planning was a satisfactory motive in work. Previously he had felt that work was only for the benefit of other people. My being able to help him develop his narcissistic action motives allowed his idealization of me to continue. He became curious about how I went about my work and about my personal motives. As his work took on added significance he began to relate differently to people in his office, more individually. He still hated them all as a group.

At this time he decided to have a child with his wife. They had a little boy. He shifted a great deal of extended narcissism to his son. During the period of his wife's pregnancy he began to collect oriental rugs, an interest that extended from a rug he admired in my office. He identified his own mind with the patterned perfection of the rugs. The narcissistic satisfaction he derived from being a husband and father was sufficient to allow him to break off therapeutic contact for financial reasons. After 2 years he returned, concerned that his narcissistic investment in his wife and son had led to overcontrol and to what his wife complained of as tyranny. Subsequent therapy with this man has achieved ever greater narcissistic development. The advantages of going through developmental periods as an adult is that development is more fully conscious and can be put to conscious use. At this point he is the father of two children, and for the first time is extending himself into a supervisory position. He is extremely philosophical about his life, realizing how much time he lost and how careful he must be about every life decision. His increased sense of goal formation makes him an exceedingly careful planner. He has also taken up weight-lifting in an extremely serious way. He wants to build up a narcissistically impenetrable shell to protect himself from the anxiety of being destroyed. It seems that he can never fully get over the narcissistic breach, the sense of broken subjective coherence that requires another person as a symbiotic partner, at least in ongoing fantasy. I fulfill that role for him intrapsychically. However, his development continues and his middle-aged integration of identity, his remodelling for the future, would appear to offer him insurance against any further psychotic disintegrations.

Nondominant Hemisphere Nonpsychotic Disorder (An Obsessive-Compulsive Case)

The first case to consider is a case of *obsessive delirium*. Obsessive delirium refers to a patient's intense involvement with obsessive doubting, guilt, worry, and novelty anxiety to the point of being unable to become involved in ordinary activities. It is a case that instructs us about the potential of some disorders to reach the stage of manic or depressive major affective illness, for this is a case that would probably have progressed to a manic or major depressive phase without treatment.

The patient is a twenty-four year old who developed a severe state of doubting and anxiety, as he was rejected by a young woman. Before involvement with her, the patient had avoided all but brief encounters with women. Throughout his whole adolescence he had quick trysts with an older woman who lived in his building. He avoided any emotional intimacy with her. The instigating relationship was a new one with a young woman who was infatuated with him for a while, but who dropped him when she saw how far he was from maturity and a sense of command of his own life.

The patient had been spending the 3 years since finishing college procrastinating, staying up until 4 a.m. watching old movies and taking occasional courses leading toward a master's degree in musicology. He had taken violin lessons from the age of eight, but with less and less belief in a musical career. It soon became apparent that he had felt that if the relationship with the young woman had developed any further, it would have alienated him from his father.

The patient's father is a severe, obsessive surgeon who has developed a ritualized compulsive routine for assessing the symmetry of all members of his family: his wife who is a vegetarian and a pianist, his daughter, who is a nurse, and his son. His mother wanted her son to become a famous performer. For that reason she had given him violin lessons ever since he was a young child.

When the patient entered treatment he was faced with the dilemma of finally doing something about his musical aspirations or abandoning them. The problem was that the patient was not able to think of himself as an adult dedicated professional. He felt too dependent on his parents. The affair with the young woman triggered a severe conflict. He was afraid he might have developed a venereal illness. He did not know if he should tell his father about the affair. He was tormented. Finally, because he was afraid that his body had been damaged, he did tell his father. His father's increased search for asymmetry was linked in his mind to the damage he believed his sexual involvement might have caused. Now his father harassed him with recriminations, hovered over him, studying him, even when he was in bed at night. He became sleepless. He began to obsess about every aspect of his relationship with the young woman wondering what had disrupted it. Was it *this*—which occurred on a particular date, or was it *that*—which occurred on another date at another time. He developed a preoccupation with thoughts of her other lovers that exacerbated his sleeplessness. He felt that his sleep had been damaged. In the 2 weeks before seeing me he saw his allergist, a urologist, an eye doctor, a psychiatrist, a psychologist, and a social worker. They all agreed that he was hypochondriacal and obsessive. How did he know who could help him?

The patient was handsome, symmetrically groomed, with a perfect part down the center of his hair. He gave childish performances, aptly, but pathetically imitating birds or animals. He had a whole repertoire of sounds. He was full of puns, developed over a lifetime. They were obvious childhood puns such as "What do you mean, do I like dates? I like figs." He had a tendency to trot out favorite puns over and over, desperately wanting to be loved and admired and to entertain.

He had a reputation for being somewhat odd. He would do anything to get attention, intruding inappropriately into others' conversations, or asking questions just to be noticed. He was a great fan of old movies, capable of imitating many of the characters. He could give dates and names of performances at the drop of a hat. The same obsessive tendency was apparent in his life account. He remembered dates, including days of the week, of many events going back to childhood.

When the patient began treatment he was literally exhausted by mental efforts, doubts, and body concerns. The idea of having an emotional life that related to life events had never occurred to him. However, when questioned on emotional issues, he would sometimes burst into song with sudden melodious beauty. The affects in the song related directly to the subjects discussed in the session. A moment after the song was begun, singing became simply a performance. Forgetting his original number, the patient would go on to others that he sang oblivious to their meaning, or to the persons in the rest of the building who might hear his performance.

Exhausted and panicky, but unaware that he was either tired or anxious, the patient was losing a sense of himself as a real person in the world. He was using up his neurotransmitters. Because he had lost his capacity to concentrate, I placed him on elavil to regulate his serotonin and norepinephrine systems and to stabilize his frantic environmental search. I described the medication as a temporary measure, not given because he was damaged, but because he had exhausted himself with ceaseless doubting. When he believed I would help him he calmed down.

The patient had hypomanic habits of intrusiveness and inflating object-self identity that showed no concern for social appropriateness. However, his denial was not complete as he was aware that he was suffering. It became apparent that he had an obsessive compulsive personality with marked dependent and hypochondriacal features, and that given the developmental chore of realizing himself as an adult, he was failing. The late adolescent identity crisis that he had postponed as long as possible confronted him now with the anxiety of leaving childhood representations behind. He felt a delirium, almost, a melange of homosexual and heterosexual doubts and inhibitions. In session, I told the patient he worried about the wrong things, that he couldn't handle the relationship with the young woman because it meant growing up, which he had been postponing as long as possible. I put his conflict in developmental terms. I told him that he had known what his father's response to his affair would be, and that he had told his father about it in order to be forced to remain at home as a child.

Nevertheless, in his therapy the patient showed enormous curiosity about life outside his household. His own whole family encouraged and allowed extreme regression. The father's obsessive compulsive body anxiety ruled the house. In response, the boy had developed an inner body sensitivity to a pitch that made up for his lack of awareness of affective life. His older sister, the nurse, was constantly making fun of his body concerns.

Thus, there was a great drama within the family. Everyone outside was "goyish" and inferior. The whole family shared a nondominant hemisphere concern for trivia and for body image analysis.

As treatment progressed, the patient began to compile a compulsive store of insignificant details about his therapist's life. It was apparent that this feature detection was animated by a desire to understand how a reality other than his own worked. He decided that he wanted to become his therapist's son, to move in with him and his family, to live life over as a child. He almost believed that he could. He made gestures to jump into his therapist's arms, lap, or to crawl up his pant leg. The transference was due, in part, to a complete impoverishment of his reality sense about the world outside his family.

After a few months the elavil was discontinued. Following some brief flirtations the patient began an affair in earnest with an older woman his sister's age who was a psychotherapist. This displacement of transference gave him an opportunity to expand his world. His parents accepted this new relationship, and his sister also began therapy.

My point, in presenting this fragment of treatment, is that we see a patient with a nondominant hemisphere disposition. He has no difficulty in loving. He can love me easily. He craves being loved himself. A developmentally mandated task of nondominant hemisphere character expansion, however, precipitated a nondominant hemisphere obsessive compulsive anxiety syndrome, with marked stranger anxiety, and exaggerated feature detection. The failure of these ego functions and of ego defenses precipitated a regression in which he began to be flooded by repetition-compulsion centered on incomplete and traumatic identifications with both parents. If the patient had not entered therapy, he was well on the way to expanding his character in the direction of an imitation of his father's character. For the secondary gain of developing an obsessional syndrome such as the father's would undoubtedly have fixed this mode in his character. To the extent that the obsessive delirium was failing, the patient showed a disposition to hypomania.

We can understand the identificatory tendencies that focus on the father and mother, his participation in movies of the past, and his imitative speech and singing patterns, his mediocre violin idiom with little original content, and his incredible sensitivity to feeling signals from the interior of his body, as all exaggerated nondominant hemisphere modes of function. The whole family shares an agoraphobic mechanism, an anxiety about leaving the family compound. The entire outside world is novel. The patient's body exists as a damaged image. He undergoes intense feature detection about his body. Every session begins with an account of the state of his body parts and orifices, and the steps he has taken to protect them. The major focus of the therapy is helping him center and identify his self, through helping him become aware of his affects and anxiety as his own possessions. To this end the patient reports dreams in each session which are very much on target and which he is capable of analyzing with ease. It is as if he had not developed defenses against direct dreams about his experience. This makes the therapist into a person who can identify him and his affects, and the analytic work of the therapy becomes a new self-reflective structure that can gradually be integrated as an adult structure of reflective, self-observing identity.

A Case of Dysthymic Disorder The patient is a forty year old woman who first consulted me when she was thirty years old. At that time she could not decide whether to work with a small group of people she knew in setting up a small publishing company or to go to work as an editor for a large, well-established company. It appeared that the friends who were setting up the company were unreliable, so she decided to take the conventional employment. Soon she lapsed into a sullen state, preferring to be silent in the sessions. This state corresponded to the way that she had felt during a great deal of her growing-up period. Her mother had been chronically depressed, seclusive, and withdrawn. Her father was a successful engineer, but he denied the emotional reality of the household and collaborated in keeping the household atmosphere dark and foreboding. The furnishings were always in disrepair, the house was dirty, and the patient felt ashamed of her home and her family.

She has a sister 3 years younger who has suffered a severe psychosis in her adult life. The sister exhibits a mixture of paranoid and depressive symptoms not unlike their mother's. When she came into treatment the patient felt guilty about her sister's illness because she felt that it was her responsibility to raise her sister in lieu of her mother. She had resented the maternal role thrust on her because of her feeling that her father was an egoist, a man concerned with his own work, rather than with her desire that he should help run the family. When she confronted him in her youth, her father would say that things would work out in the future, which they did not.

The early transference was conditioned by her experience with her father. For years she would not really trust me. When the chips were down and she felt some immediate need for help there would be sessions when a real feeling of closeness and mutual respect would spring up, but this was not usual. Her attitude also reflected her family's distrust of outsiders. For years she did not ascertain that I could understand her Jewish customs. Being Jewish had compensated her for her sense of an empty family life. Both parents had come from extended families. Each felt an outcast from these families, but each parent still felt that they had a more important home and family outside of the nuclear family. The patient felt that her therapy, like her family, was never going to provide her with what she needed.

Her suffering also was involved with the feeling that things would never work out. When her work achievements were not recognized by an older male boss, she suffered and simmered for weeks on end. During these periods her work was excellent. She had developed the ability to perceive reality sharply. Her response to the massive denial in her family was to develop a fine sense of what was really happening. With laborious and painful zeal she would approach her editorial work, always ending up by organizing a mass of material in an extremely logical way. She also had a fine sense of humor, a sense of what was bizarre or unrealistic in the world, and she enjoyed sharing that with other people, particularly when she felt expansive.

This patient developed her nondominant hemisphere identity functions to a high degree. By going against the trend of denial in her family, she sharpened her sense of reality. She had a strong moral sense that was always pitted against other people's selfish actions. She was not quite a moral masochist, as she did not espouse the moral side in a superior fashion; rather, she upheld it as a logical regulator of reality. Thus, a great deal of her upper level identity structure, her personality, was vested in producing realistic versions of a conceptual morality. She was concerned with the real inculcative value of the textbooks she worked on. Thus she identified her own object-self social function with a portion of society in an appropriate way. However, when she felt depressed, she felt that the textbooks had no intrinsic value and that they existed only as a form of commercialism that she abetted. Pondering this deepened her insight into moral and social conditions. She compared the present with Victorian times. Her sense of the literary past, especially as expressed by woman writers, was sound and interesting.

For leisure she read detective stories; she liked solving mysteries and doing erudite crossword puzzles. She liked to knit afghans; she enjoyed the deliberate movements and the deliberate emotional life of her four cats. Thus her feature detection and recognition of pattern gave her nondominant hemisphere engendered pleasure.

This patient's character structure was logical and morally obsessive. She regularly emerged from depressive periods with some burst of humor, like the sun shining after 40 days of rain. In helping this patient with her depressions the most important factor was the therapist's sense of not giving up, despite repeated invitations to do so. For she recurrently felt she would never have a good relationship with a man, that she would never have a moral boss, and that her life would never achieve any significance.

The patient's relative lack of involvement with her dominant hemisphere functions had tangible effects on her life. She sensed that she was not creative. She could not devise goals and attempt to reach them in her work. She felt that she had no vision about what would be important in the future. She castigated herself for the lack of these qualities. She, indeed, had no master plan in her life, nor could she seem to develop one. Gradually, much of the work of the therapy came to focus on why she could not develop goals herself. The whole burden of her survival had depended on her being alert to her environment, to her mother's intense and sudden angers and her attempts to stave them off. Like other depressives, she is unable to withhold empathy from fellow sufferers.

The patient developed an intensely masochistic relationship with a man. She was deeply concerned with his pain, and unable to monitor the progress of their relationship. He took advantage of her, insisting that she be there for him every moment, while he would go about his own business when he felt like it. He was not responsible with her financially. After a period of 6 years this relationship clarified and became viable. The man is a computer specialist. She finally gave up her editing work, which indeed was very other-oriented, and learned computer operations herself. She was easily able to identify herself with the computer type of thinking. In fact she was quite curious about the analogies between computer operations and human mental operations.

The relationship with the man took on a more vibrant quality during a period in which he admitted he was probably dying of untreated ulcerative colitis. For years he had denied the seriousness of the undiagnosed condition, by taking massive doses of marihuana to detach himself from the pain. At this point he had lost about 40 pounds and was becoming physically weak. Finally the patient dreamt that she herself was dying. In therapy I suggested that she has represented her future loss in this dream. It seemed to me that the patient's loss was *his* dying. This realization allowed her to really help him for the first time. The patient's lover got medical help after decades of pain. In his relief and acknowledged sense of connectedness this man was able to reconnect himself to his own past. He had a sixteen year old daughter from a brief earlier marriage, whom he had not seen for 12 years since his wife had left him. He had not paid any taxes or filed with the IRS since that time. Finally he faced the music. Together, the patient and her lover began to open to the extended social world.

He became financially responsible, no longer spending most of his money on marihuana. She decided to have surgical repair on her uterus rather than the impending hysterectomy for serious fibroids. The patient and her lover decided to marry, and possibly at this late date to try to have a baby. The patient feels a more continuous sense of participating in her life. One further change the patient made was to give up her editorial work and to seek retraining in the computer field. This way, engaged in the same work as her future husband, she feels they may eventually be free to make their own future without pressure from other bosses.

In essence then, the direction of the therapy was working with the nondominant hemisphere. The patient's dysthymic depression was alleviated as she felt that she was involved with life, that her relationship was pre-eminent and reliable, and that she could make plans that would include her emotional needs. The depression may be seen as a failure to feel participation in her life as a process. Instead of feeling that her life has no future and that nothing can work out, she now feels that her life is unknown, but that it is deeply interesting and that the process of working out emotional themes will give her life direction. She does not know if she will have a baby at this late date, and she can accept the life that follows in either scenario. She accepts the help that she received in therapy, thus resolving the father transference. Her own antidenial reality testing mechanism is now an essential, valued part of her character. She values her ability to do detective work in her computer analysis and does not rule out the possibility of combining this with her editorial skills, skills that had brought her to the position of managing editor.

A Case of Masochism

The patient is a forty-five year old woman who has endured a very masochistic life. She is a warm, outgoing person who remarried about 5 years before but found that her romantic expectations were not being fulfilled. Instead, her health had deteriorated. She had possible rheumatic arthritis, with other connective tissue involvement and indications of peripheral vascular Raynaud's disease. She had intercurrent polycythemia. She had several operations on her feet that had increased rather than decreased pain. In short, she was on her way to becoming an invalid. She had not worked since her second marriage. The patient relied implicitly on various doctors, one in particular having become her personal counsellor, even evincing sexual interest in her. This doctor was the reincarnation—as far as she was concerned—of her beloved father who had died many years earlier. She entered therapy in a state of painful emotional disarray.

The patient had grown up the youngest of three daughters. She characterized her oldest sister as an evil genius who tyrannized over the family and completely preoccupied the parents. Her insight into her sister's motives was acknowledged by the sister who called the patient a "witch." The mother protected the older sister from the father's wrath. The mother exercised profound denial, believing that peace should be kept at all costs. The patient's role was that of peacekeeper. She would sacrifice herself and her own interests in order to fulfill the mother's mandate to keep peace. When her beloved father would get angry his asthma would flare up, and she feared he would die. When he would start to get angry he would seclude himself and the patient would go to him and try to soothe him. Thus keeping the peace came to entail a romantically tender attachment to the father that had to be concealed for fear of inflaming her older sister.

As a child the patient was always being the watchful observer. She would never feel fully involved in life herself. She would take satisfaction instead in predicting what could happen and trying to make things come out right. The one control of events she had was to sacrifice herself. She had developed this mechanism to a high degree. For instance, she was concerned that doctors receive every recognition and compensation. She would even submit to a dubious operation if her doctor recommended it. In her own family life she was able to help everyone get ahead but herself. She helped to manipulate all her children into good educational situations, and then into good job situations. She would become so active in helping people, and so self-sacrificing, that in the end other people felt so guilty they couldn't stand being near her.

The therapy first focused on her wish to become an invalid in order to get paid back for all the help she had given others without receiving any herself. This line of interpretation, as well as some expected deeply positive father transference to me, resulted in her going back to work. In her work, she soon took over many functions not properly hers, and with more efficiency than they had ever been done before, exposing her boss' weaknesses. Consequently, her boss first tried to exploit her, and then, when she showed him up, tried to fire her. Analogously, her first husband beat her, and she endured his abuse. She continued to try to build up his career by helping him with his work and using family connections to increase his business. She felt that he would ultimately appreciate her goodness. She did not want to break up the home and upset her two daughters. Instead, she put up with her husband's beatings for several years. Finally, as his sadism mounted, she entered into a divorce agreement (manipulated into using her husband's lawyer), in which she received minimal child support, which she had to supplement by her own efforts. She showed deep love and affection and appropriate concern for the children.

The pattern was clear. This was a woman who had developed a masochistic character structure. The essence of that structure was that she diminished the self function and the object-self function and inflated the object's function. She could endure physical pain without a loss of concentration. She lived vicariously in the present and maintained fantasies of compensation in the future. As we investigated this nondominant hemisphere mechanism of overidentification, the patient began to express more depressive affect. She felt suicidal and revealed a deep fantasy of joining her father in heaven. She expressed the fantasy that if she acted in her own interests and went against her mother's advice, she would kill her mother. In fact, that is what her mother would always tell her when the patient complained to her about her life with each husband. The mother would say, "You're killing me."

Finally the therapy came to devolve upon the fact that her new husband's two children had gotten wonderful jobs with the patient's help. One of her own daughters had married, and the other was working. She and her husband did not have an extra penny, but the husband's children did not contribute to the household. As a matter of fact, the patient and her husband slept in the living room. She had no privacy. The bathroom, telephone, and bedrooms "belonged" to her husband's grown children. She would work, standing on her painful feet, and be expected to come home and cook dinner—a different menu for each of the children, and then drop into bed exhausted as her husband reaped the gratitude of his children while she lay in bed listening to their laughter. Finally the husband and his two children began to avoid her. The husband denied his every communication with them. The patient continued to turn her paycheck over to her husband.

Currently, the patient realizes that the only out for her is to take a stand. But she is so embroiled in her masochism that she does not know how to do so. This is a case, then, of a nondominant hemisphere characterological masochism, with psychosomatic illnesses that feed secondary gain into the masochistic character patterns and family expectations. The character structure is distorted in such a fashion that affect response is mainly through suffering that invites guilt. Surely the guilt is responsible for the social framework that helps develop and also maintains the character structure. Thus, as a family structure mechanism of nondominant hemisphere social formation, the guilt insures that some family members prosper during difficult times in reality, while others, who sacrifice, only prosper when times are good.

The patient feels a chronically low self-esteem. This acted as an inhibiting factor when throughout her adolescence young men were attracted to her. She could not believe that it was really possible that she was being courted. At this point in her therapy the so-called negative therapeutic reaction is very strong. Pointing out her self-sacrifice disabuses her of a lifelong, and still strong oedipally fixated fantasy of eventual reward from her father. Her self-esteem requires sacrifice. She cannot easily absorb information about her sacrifice and interpretations in this area. Instead, she pushes doggedly onward in her litany of abuse. When she is confronted more forcibly about her role in arranging her own sufferings, then her fantasy of obliterating the objects who take advantage of her is exposed. This increases her guilt. Thus, guilt, low self-esteem, and suicidal feelings directed to the object-self intensify as the masochism is challenged.

It appears that only through the therapist becoming a transference figure for the recompensating father object can structural changes in the identity functions of the nondominant hemisphere be undertaken. It is a mistake to analyze too much of this so-called positive father transference too soon, for the underlying ambivalence is at its most unconsciously intense organization here. She must have some ongoing mutative experiences to show another way to participate in life as part of the ongoing therapeutic experience before the father transference can be thoroughly analyzed.

The patient tries to find evidence that I am actually related to her. She felt "sure" that a rug in my office was originally in her own house when she was a child. She also had a strong wish to give me her father's books on the mystical Judaic laws, which she believed might have great fundamental importance. If she were certain that I was indeed her father, then she would be liable to psychosis. When I interpreted her wish to give everything to me including all of her love, as I refused the gift of the books, she felt suicidal, as if her love were being turned away. This also evoked powerful feelings and fantasies of being reunited with her father in death.

A Case of Dominant Hemisphere Triggered Hysteria

The patient is a forty year old woman who originally entered therapy when she was twenty-six. At that time she suffered severe feelings of guilt that had been triggered from two sources. First, she had refused to allow her new husband's crippled younger brother to come to join them soon after their honeymoon. Her husband had never mentioned this during the courtship. The younger brother, who had been attached to his older brother as a substitute father, promptly drowned himself in a swimming pool. The other source of guilt was the fact that the patient's younger brother had just gone into a state hospital for an apparent schizophrenic episode. The patient had insisted he enter the hospital. She had been partially responsible for raising this sibling as both her mother and her stepfather worked as she was growing up.

The past history of this patient was dramatic. When her mother was a beautiful sixteen year old, she was wooed away from the fishing village of her origin by a foreign sailor, who turned out to have syphilis. When the patient was just a year and a half old, a first son born to the couple died of congenital syphilis. At the same time the father died of tertiary syphilis. This left the mother, now nineteen, with her infant daughter in America, a foreign country, somewhere in Illinois, in the winter, on a flower farm. They were penniless and starving, and the mother could not speak the language. She had to leave her infant in the crib with the icicles on it during the day while she foraged for some sustenance. Finally the mother learned of some relatives on the West coast, whom she joined with her baby daughter. But the price of their hospitality was that she become a servant to these relatives.

Finally another man fell in love with the mother, who was still a young woman. The patient was four when her mother and stepfather married. She was supposed to show great gratitude to this new saviour. He began to come in to her at night and to fondle her genitals. When she complained, he laughed at her. The patient felt sexually stimulated, but let down when her stepfather left her, for he only aroused her, never had intercourse with her nor offered her his penis. She was a plaything whom he enjoyed tantalizing. She felt that the curse of her father was upon her when she heard the relatives whispering about her and her family. She fantasized that her stepfather was investigating her, searching for the ravages of syphilis that had killed her father and her brother. She felt that any sexual response—and she did feel sexual response—would lead to her own death.

After her parents had two boys, five and eight years younger than she, it was her responsibility to take care of them. At times she masturbated and felt guilty and damaged. She believed she was ugly; her mother kept her in shabby clothing with her straight hair tied back severely. She was tall and lanky, and felt ungainly and "poor." She developed a fantasy of a rich, beautiful blonde girl, socially poised and graceful, her apotheosis.

All of these feelings haunted her in her adult life. She could not have an orgasm because that would prove her a "pervert" like her father—her father's daughter. She was afraid of the man's genitals damaging her. Her work as a nurse was stimulating, and she could help "damaged" people. But she did not like it because she felt like a personal servant. She was resentful of female colleagues. She felt that, like her mother, they did not want her to do well. She could not allow herself to win in competitions with females.

Her husband, like her stepfather, had no sensitivity to women except for his own mother. He insisted on dominating her completely and taking her sexually when he felt like it without regard for her satisfaction. He blamed her for the death of his brother. These forces led to her divorce soon after the beginning of her therapy.

After several years of therapy, with interpretations about the above material, especially genetic interpretations that tied her past experience to her reading of the present, the patient made some progress. The essential oedipal interpretation stressed that she could not allow herself either to compete with another woman or to feel sexual pleasure, because she felt guilty as a survivor, as if she too might become a casualty if she relaxed for a second. She also felt guilty because of her sexual response to her stepfather. Although she longed to complete the sexual act with him, this would mean obliterating her mother who had already sacrificed in her own life so that the patient might live. Partially overcoming the oedipal obstacles, the patient became a supervisor in her work, with other woman working under her direction. She had sexual relations with men in which she could finally have an orgasm, albeit only during oral or physical stimulation aside from intercourse. She stopped therapy at this point.

She returned when she met a man with whom she fell in love. He had five children from a previous marriage, between ten and eighteen years old. After they married, his former wife developed an operable cancer, but she decompensated emotionally, and was unable to continue to care for the children. At this point, when the younger children were twelve and thirteen, it was necessary for the patient to take them into her house and assume the mothering role. As she tended them, some of the effects of their neglect began to be overcome. Her husband, feeling less guilty, began to do better in his business, so the patient no longer had to work in her previous job. Surprisingly, the patient revealed a fantasy in her therapy that she had harbored a long time. She had been a fine arts major in college, and she wanted to become a fashion designer.

With hardly more than a sense of permission, and while mothering the children, the patient became a fashion designer. In only a few years her career blossomed to the point that her designs were known throughout the United States and Europe. Some of the dynamics of her fixation were involved in this success. In her determination to overcome the sense of being poor and damaged, she perfected costumes that sublimed her originally poor body image while they represented integration into monied society together with an aesthetic relation to life.

The present case illustrates the power of fixations over the subjective side of life. Originally fixated during the beginning of the primal scene period of life, the patient developed an exaggerated sense of her own powers of survival. She felt that she had created herself, and that she had taken over her brother's and her father's identities. She subjectively encompassed them. Her seduction by her stepfather re-enforced her sense of herself as extraordinary and fixated her to oedipal masturbation. Having been forced to wait for sexual gratification that never arrived produced fixation together with anxiety about her inability to invent an action plan that would yield gratification.

Because of her stepfather's pederasty, and because of her dead father's sexual reputation, she came to identify sexual activity with maleness. The patient was left with a strong wish to recreate herself in a masculine form, to "complete" the state of her genitals. In order to achieve gratification she sublimated the oedipal fantasies into artistic work, 16 hours a day, if necessary, to bring subjectively conceived plans to fruition. A continued, imperfectly internalized identification with her father and her brother (both of whom died of syphilis) still makes it difficult for this patient to achieve satisfaction, for she fears that her own satisfaction, if actively sought, will lead to death and destruction of others whom she loves.

A Case of Nondominant Hemisphere Organized Borderline Character Disorder

The borderline character disorder is a lifelong disposition to process life experience in a distorted, predominantly nondominant hemisphere way. The extent of this major distortion suggests that borderline syndrome arises early in life, and this in turn suggests that cases of borderline disorder can teach us a great deal about the development of hierarchical functions in the nondominant hemisphere.

The patient is a woman who entered therapy with an analyst (not myself) at the age of eighteen. She was away from home at college. As graduation approached, and termination from therapy with it, the patient began to realize her complete aloneness; she felt suicidal. Her therapist sent her for hospital treatment. This hospitalization cut her off from her therapist in a sudden and painful way, for termination was not worked through. Instead, her therapist went on vacation, after which he didn't see the patient again. At the end of her stay in the hospital the patient returned to her home town. There she sensed her mother's continuing hatred for her. The patient then made a serious suicide attempt by taking a large overdose of barbiturates that rendered her comatose. She was hospitalized in a long-term psychiatric unit for more than 2 years. When she left the hospital she continued the therapy begun in the hospital with her inpatient therapist until the therapist's sudden death several years later when he was on vacation.

During her 2-year hospital stay the patient manifested severe major depression, without psychosis, but with a borderline personality organization that included a real deficiency in her sense of her own objective development as a person. She maintained a nondelusional, but primitive organization of her consciousness, that she referred to as a "right", a "left", and a "center" portion of her identity. The right was herself as a little girl, the left was her sense of herself as identical with her mother, and the center was a modulating influence that she felt was closest to her real self. The splitting in her experience extended to her body image. She felt that her body was unreal and that it did not truly belong to her. She realized these conclusions were inappropriate, but nevertheless felt them to be true. The patient's left identity corresponded to a sense of complete identification with her mother.

When she felt very angry, like obliterating her mother, and when she once developed a compelling image of stabbing her mother with a knife, she felt as if she were exactly the same as her mother. She experienced the sameness as a strong left-sided sensation in her body. The "right" identity as a little girl encompassed the more satisfying period before her sister was born when the patient was four. At the same time she realized how much her mother hated her, and this realization and the oedipal disappointment with her sister who became a burden to her, promoted a traumatic breach with her previous identity.

I took over this patient's treatment after her therapist died, since I had been covering his practice when he was on vacation. Despite the fact that her options were carefully explained, the patient felt that she had no choice but to continue with me. Although she had clearly developed a deep affection for her past therapist, she could never acknowledge it. For years she dreamed, and sometimes felt, that he had not really died, that his death was not real, but merely a ruse to get rid of her. In this respect she felt that her mother was right, that she herself was really unfit for any human companionship.

Maternal trauma was at the root of this patient's problems. Her mother disliked her openly from the beginning of her life. She made it clear that she wished the patient had never been born. She often went into unpredictable rages at the patient, "turning black as a cloud" and striking her. The daughter responded by becoming as quiet and unobtrusive as possible. She attempted to "perfect" herself by making no demands on her mother. The patient was the second of three children. She had a brother 2 years older than herself. He was her "hero." In fact, when he became an adolescent and abandoned her in his adolescent quest for girls, she got a dog whom she called "Hero." She consciously bought the dog as a replacement for her brother.

When the patient was four her sister was born. The patient consciously wished for the death of her sister as she observed her mother treating the sister with more care than she herself received. As an adult the patient remembered feeling at the age of five that she herself could be seen through, and that her death wishes toward the sister were the cause of her mother's rejection of her. At this point she split the image of her mother into an unreal idealized version, based on a picture she had seen of the mother before the mother was married, and the real mother who was permanently bad. She identified the previous good image of herself, now lost, with the lost image of the good mother, an image of the mother before she had met the patient's father who had unaccountably spoiled her by marrying her. As time went on the little girl tried to redeem herself by stilling all her demands on her mother, and all her desires for her father. She felt that her mother's severe dependency had been the fault of her father. Her father had, in addition to the three females at home to support, two maiden aunts, a retarded brother, and his own mother to support, which was why he was always working. The mother had been a friend to the father's sister before she met the father. This sister became emotionally ill, and represented to the mother invidious claims made on her husband. The patient, when she was born, was seen by her mother as the figure of the mother's own sister with whom she competed unsuccessfully, and also of the father's sister who made such financial demands.

The patient's father was viewed as a powerful and morally superior man. By working 14 hours a day to support all his women the father proved his superiority. He never asked anything for himself. This demeanor also brought him to a position of financial prowess and influence in the banking community. However, his own money was always tied up in contingency deals that were tied in to future acquisitions, and thus there was always some immediate financial threat, coupled with the promise of future ease.

The girl patterned her own morality and intellectual approach to life on her father's. She developed a rigid moral system. She always attempted to go by the rules, but usually found that the rules let her down. She was an expert at uncovering hypocrisy.

The patient's father did little for her emotional health. During adolescence, while the patient was trying to develop her social life, she realized that she had a strong desire to kill her mother during one of her mother's rages at her. She had wanted to pick up a knife and stab her. At the same period her brother had withdrawn from her. The other girls in her school were drawing away from her, leaving her behind as they pursued boyfriends. After the incident of wishing to kill her mother she withdrew into a depression and stopped her school work. Her father did not appear to notice her depression, and she blamed him for this abandonment. She took long walks on the beach with her dog, feeling suicidal, and protecting herself from impulses through a kind of mental confusion.

During this period the patient developed severe acne, possibly due to stress-related interference with her hormonal development. She also failed to menstruate regularly. The patient's mother took an intense interest in her acne, as she had on some occasions become attentive when the patient had been sick. Her mother brought her for prolonged sessions to a dermatologist. She felt the sessions to be extremely painful and bitterly humiliating. After about a year of these sessions her problem was worse rather than better. Her mother withdrew from the preoccupation with the acne. The dermatology sessions later became a transference image, for the patient felt that coming to the sessions and exposing herself to the probing questions of the therapist was like the sexual relations she had attempted—painful and humiliating—and just as effective as seeing the dermatologist. It had become clear to the girl as an adolescent that her mother wanted her to try to become pretty so that she might meet a man and leave the home. Indeed, the mother's rage reactions did abate after the girl had gone away to college.

When the father was at home he preferred to talk to the patient. She was like a protégé to him. He talked to her about his business dealings as he might talk to a colleague. She felt embarrassed, first because her mother was jealous of her in this role, and also because the mother was jealous of her high intelligence. However, the patient felt that she could not understand what her father was talking about. This amounted to an intellectual inhibition. She did not feel she could tell her father that it was hard to pay attention to him, so she endured the sessions. This also figured into the transference relations with me. Although she understood interpretations she could not use them completely, as they remained somewhat obscure to her in an emotional sense.

Therapy with this patient was difficult, for there was hardly ever any sense of emotional warmth in the relationship. The patient's main form of transference, which had, of course, to be worked through before termination, was that which derived from her relationship to her mother. She felt her link to life itself existed in her relationship to me; that without me she could not survive, or that she would end up as a "shopping bag lady." As soon as she would voice these sentiments she would express a feeling that she couldn't stand me, saying that she never could, and that she had no choice in having me for a therapist. Pointing out to her that these were feelings engendered in her first with her mother induced in her the wish to find some other person to make up to her what she had missed with her mother.

As therapy progressed, the feeling of dislike for me became stronger and stronger. I attempted to interpret the dichotomy between her dependency and dislike by telling her that she was trying to renounce potential feelings of love for me out of fear that she would lose me if she realized these feelings. I added that she feared that I would leave her as her first therapist had, or die as her second had. The patient acknowledged these fears, but she corrected me, saying that she had never loved either of these therapists, and that she also would never love me. She added that she needed me in order to connect with her life, but that this was no substitute for the close relationship with a woman that she had always missed.

Because of her identification model deficits, i.e., her negative trauma, she was identification-prone. Identifying with me as the analyst, the patient became proficient at self-analysis and at analyzing other people's character structure. For 6 years after coming into treatment with me she continued to be unable to work. Her self-esteem was so low that it was too painful for her to cooperate with other people for prolonged periods of time. She lived in an apartment paid for by her parents. Finally, she got a job selling advertising space on the phone. She rented a loft. At this point, the loft became her whole life. Building the loft was clearly and consciously equivalent to building her own object-self structure. Until it was completed, and perfected, she did not feel she could have anyone visit her. She also felt guilty about any addition she made to the loft, because she felt that she didn't deserve it, and because she didn't feel that the money to pay for it was her own. She also felt guilty about the sessions with me that her father was paying for, insisting on cutting down from four to three. Every move she made in the world had to be morally weighed.

After she developed a sense of integrated object-self identity, as reflected in her completion of the loft space, she began to try to perfect her body. She had grown somewhat obese. Joining O.A. (Overeaters' Anonymous), she went on a strict diet. The organization's rigid rules suited her sense of morality, and she used their rules to begin to loosen her own inflexible morality. Her sponsor became an important person to her, someone she granted an overriding ability to know what was good for her. Hunger for a real mother increased. She sought out her own mother after she had lost a lot of weight. The mother enthusiastically helped her to shop. Now in her middle thirties, she bought clothes that were appropriate for a teenager, and she was again crushed when she attempted to meet people her own age and found herself uniformly rejected. She felt that she did not know how to get along with people her own age. It reminded her of adolescence, when she had had only one friend in her life. As an adolescent, she recalled, she had been amazed when she had visited her friend's house and found there a mother who was warm and responsive, and interested not only in her own daughter's life but hers as well. Eventually that friend had casually dropped her.

As the patient re-experienced social rejection, she felt disconsolate again. Furthermore, the mother returned to her typical coldness as she sensed the daughter's new failure. At this point the patient felt more deeply depressed. She lost interest in the loft, in the brief, semisuccessful dating she had begun, and in O.A. In a search for compensation she moved out of the loft into a living situation where she occupied a space in a woman's apartment. As she gradually realized that she had chosen a woman precisely like her mother, she retreated into her room much as she had done in her middle adolescent depression.

However, she kept trying, despite feelings of hopelessness. She went to an organization that helped women who wanted to enter the business world and took courses in selling. Finally she found a kind of work that she had an affinity for, selling business machines. She had a remarkable memory for details and good organizational abilities that allowed her to learn everything about all of the competing machines. After about 3 years of devoting herself to this work, she became financially independent and self-sufficient. Despite the fact that her father could support her and the therapy, she felt that she had been draining him. As she began to make more money and to have some insurance available to pay for some sessions on her own, she insisted again on cutting down one more session. Her sense of independence was involved both in being able to pay for the sessions and in being able to forego the sessions. She drove a hard bargain about fees. In her work she also drove a hard bargain, but she prospered as she gained a reputation for being absolutely fair, to the point of filling an order even if it meant that she would lose a great deal of money because of a mistake she had made.

When her financial life was in order the patient entered a psychotherapy group with a woman whom she soon loved and idealized. With this woman she had an experience of being loved and accepted, which she had never known with her mother, or with me. Once again, the money became the means of cutting down the sessions, as she felt she could not afford to continue with me and also see her group therapist. She also brought up the feeling that she was now near forty years old, that she had been in therapy for about 22 years, and that she wanted to see if she could live without it. Under these pressures we set a termination date several months ahead.

The latter part of her therapy dealt with her symbiotic feelings about ending the treatment. She felt that ending the treatment was tantamount to killing me, yet she felt that she had to be concerned about herself. At the end of therapy she had no affection for me, just respect, and a profound identification with my conceptual assessment of reality.

The termination phase constantly and painfully worked through the transference. The patient would drive me away from her forcefully, and then relent, and admit that she felt that she couldn't survive without me. I would interpret the mother transference and ask her why she couldn't experience both the need for me and the wish to destroy me at the same time. I told her that it was because she couldn't stand the loss of self that was identified with me, a self that was modelled after her identification with her father. After countless variations on this theme, the patient convinced me that she didn't need me and would be better off without me. Because I was convinced, she and I felt that she left through her own efforts.

Consideration of this case shows a patient who was unable to build up her own objective structures of identity. Her personality integration was very weak, leaving her vulnerable to severe depressions. When she began seriously to think of leaving me, 3 years before termination, she experienced severe agoraphobic anxiety, re-enforced by fantasies of growing old, weak, and helpless, all by herself. During one period, her agoraphobic symptoms were accompanied by early morning waking and fluctuations in appetite. Because of this I placed her on tricyclic antidepressant medication. This produced a sense of mental integration that surpassed what she felt that she had achieved through her own or my efforts. She responded (1) by blaming me for not having given her this medication before, (2) by feeling that therapy was not really useful, and (3) by feeling that now she needed me more than ever. These feelings all had to be worked out within the context of the basic mother transference and deprivation. She stopped medication after several months, long before termination.

This is a woman who could not build up a sense of herself as a woman because her mother had not been available as a model of identification. The patient was right to complain that she could not get enough of that from me. As a child growing up in a hostile maternal environment she had needed protection from her father, as she later needed it from her therapist, but the sense of protection offered did not produce enough identifcation to help her develop the self structure that was lacking. The patient did show an exaggerated nondominant hemisphere tendency toward the formation of identifications. These did not produce false object-self structure in the therapy identifications with me, but rather real object-self structure that she experienced as her own, but incompletely integrated within the framework of her personality. Gradually she organized the identification with me and with her other therapists according to her negative oedipal identification with her father. She became a canny and reliable business person.

The patient's unusual organizational skills combined with a good perceptivity about space and the utility of objects. Her construction of her loft revealed a highly developed architectural skill. All her skills derive from nondominant hemisphere functions. We can presume, then, that this is a woman who had a genetic tendency to right hemisphere processing, and that she suffered more than most from a traumatic deficit of identificatory object availability for modelling and building the structure of identity in this hemisphere. The patient developed classical major depressions. We can see that she was more vulnerable to agoraphobic anxiety because of the fact that her object world was so depleted in the first place. She had no one with whom she could identify, and so terminate her depressions. The dearth of objects was a factor in her failure to modify the primitive mechanism of splitting that she employed in her object relations.

Rebuilding her life from the neuropsychological level, at which she was innately talented and intelligent, the patient put together an organizational and utilitarian construction of the world. Extremely logical and organized, she was able to perceive other persons' needs for salience, and thus she could find appropriate machines to fill their needs.

A Case of Acute Paranoia with Dual Hemispheric Phenomenology

The following case report is instructive in a number of regards. First, it shows the effect of a social atmosphere of paranoia in precipitating an acute paranoid episode in a man vulnerable to paranoia. Secondly, it shows the bihemispheric nature of the process of symptom formation in paranoia. There is both a strong depressive component that is averted through the paranoia and also a strong schizophrenic-like element in the triggering process and maintenance of the acute entrance to the paranoid syndrome. Thirdly, it is possible to follow the sequential events in a rather continuous way because the patient is a good reporter, and because the patient was well known to the therapist due to years of earlier work with the patient. Finally, a re-examination of a traumatic episode that occurred when the patient was eight years old that determined the patient's vulnerability to paranoia during his latency establishment of an extended social world shows how repetition-compulsion can activate a defensive and regressive and disintegrative destruction of the authenticity of the social world in paranoia.

I first saw this patient when he was seventeen. I had just finished my psychiatric residency. He was my first private case, a sufficiently florid one.

Having dropped out of school in the ninth grade, which was the first major sign of a progressive withdrawal, this patient felt completely helpless when he came to treatment. He contemplated suicide. He spent his time in his room, pacing, imagining taking a violent revenge on his older brother who had teased him and persecuted him during his whole life. When he tried to contemplate putting a knife into his brother, he turned up suicidal images.

The patient came from an interesting family. His mother and his mother's older sister had competed intensely all their lives. His mother had an affair early in her teen years, producing a son. She had been shamed and removed from the competition as the loser. Her son was sent away to foster care, and for years the family pretended that he did not exist. The patient's mother married a man who had obvious low self-esteem, around the same time that her sister was married. Her secret of having a son was kept until after the marriage, and in return the husband's secret of having had a nervous breakdown earlier was also kept a secret until after the marriage. After the marriage the competition was renewed. Each of the three times that the older sister got pregnant, the patient's mother did so within a few months. It then turned out that each family had three sons in which the older sister's sons were each a little older than their comparable cousin. The patient was a middle child. He and his counterpart were the smartest in each sibship, the most sensitive, and the ones in whom the mothers invested themselves most deeply.

When the patient was eight years old his unknown half sib graduated from foster care and came to live with the family. The children were told that this half-brother was the result of the mother's previous brief marriage. In revenge for his own desertion the half-brother returned like a spectre, humiliating the older brother in particular. The older brother in turn passed the vengeance on to the patient. Family dinners were a nightmare in which the half-brother bullyingly dominated the scene, while the father was shown as a weakling, and the patient, who was slight like his father, tried to stick up for his father. Finally, the father had a psychotic episode requiring hospitalization. In the aftermath of the hospitalization the family turned to the mother's sister and her family for financial help. In a dramatic episode with both families present, the aunt proceeded to denigrate the patient's father. Suddenly the eight year old turned on her and accused her of always putting down not only his father and mother but the whole family as well. In response to this challenge to her authority she turned the full force of her fury on the slight boy. But he stood his ground. In fact, he swore at his aunt, accusing her of ruining his family. As the whole family stood there, his aunt marched him into the bathroom and washed his mouth out with soap, forcing a good deal of the bar of soap down his throat in the process. Since his family did not intervene, the young boy made an inner decision that he would remain stubborn inside himself, where nobody could reach him. So, in a vivid rage and full of intense humiliation under real persecution the boy began a process of withdrawal that eventuated in his failing school. He had no friends, and he returned daily from school to the torment in his house. By the time he dropped out of school his half-brother had left the home to go to "reform school."

When he began therapy the patient was a small and frail, tortured-appearing youth with severe acne. He could not defend himself against his older brother's accusation that he was a "fag." Indeed, he felt drawn to his counterpart cousin who did show homosexual inclinations. He felt that no girl could possibly ever look twice at him, except in derision. In that first therapy the patient identified with me; he returned to school, he went to college, and he met a girl whom he married. He left treatment at the point that he went off to a prestigious graduate school on a scholarship to study political science.

Thirteen years later the patient began therapy with me again under the following circumstances. He had been working as a college librarian in a small academic college town. He had, since I had seen him last, failed to complete graduate school because of difficulties and conflicts about expressing himself in the area of political science. His overdetermined interest in the power structure was not sufficiently analyzed to allow him a free academic interest. After this failure he succeeded in a second career in library science. During the period of turmoil his relationship with his wife deteriorated and he was divorced. With some difficulty, and a brief consultative supportive contact with me, he had found another wife who worked within the same library system that he did. He bought a house with his new wife and had invited his now elderly parents to live in an apartment near him, which they did. This arrangement worked out well for 2 years, until a political turmoil broke out.

This was a time of waning power, money, and prestige at many smaller colleges. A new president had taken over the college and he wanted to bring library services under his own academic jurisdiction. One point of contention was the fact that librarians achieved academic tenure in competition with college teachers. In fact the patient was up for tenure at this time. Sensing the change in the power structure, the patient's supervisor and the chief librarian both went about finding themselves new positions. In this vacuum the other librarians wanted the patient to assert himself and to take on a position of leadership. Although speaking plainly for the librarians' rights, the patient did not want an advancement. Instead, one of his peers became his supervisor. Under her leadership all of the librarians agreed to hire an outsider to direct the library. At this point the president of the university began an investigation, charging the library with "discrimination" in their selection. The patient, knowing this was untrue, stated that he could not participate in such an investigation. The appointment of a new director did not go through.

The patient's supervisor now decided that she had to look after her own career. She said that she would have to cooperate with the academic president, who had by now increased the scope of the investigation of the library. The investigation became so acrimonious that the librarians could not proceed with their ordinary work. The library staff as well as the college faculty were split into two factions. The selection committee for a new chief librarian, which included the patient, rejected a Spanish applicant who was clearly unqualified for the job, as pointed out by the patient, and chose a "white" man instead. At that point the patient found himself lauded by the racist group on campus and rebuked by some of his old friends.

The critical faculty group, including the president, then enlarged the scope of the investigation going to the state level. During this third period of investigation the patient could no longer maintain his social and personal equanimity. The investigation had lasted for 2 years, and the various acrimonious charges and the various investigations began to blur and become ambiguous in the patient's mind. During the third investigation a black chief librarian was hired, a man who liked the patient and whom the patient supported. However, by this time the patient was mired in psychosis.

The patient resigned his already tenured position and reapplied to graduate school in political science. But as soon as he went to graduate school he felt that everyone there was talking about him, and he could not study. He felt that a conspiracy against him had become statewide. Jobless, with diminishing money, in danger of losing his house, and beginning to doubt his wife's affection, he looked me up again.

The Development of Symptoms For more than a year and one-half the repeated investigations subjected the patient to increasing pressure. He began to have difficulty in falling asleep; he was constantly preoccupied with his social role and with being misunderstood. On one occasion, as he spoke out for his fellow librarians, he began to feel that he was developing an uncontrollable rage. He was more than reminded of his standing up to his aunt. He felt that his own library was a family that was being humiliated. Then he began to feel that his supervisor who had gone over to the other side to protect herself had become involved in a conspiratorial way with the college president. She began to remind him of his aunt, particularly when she began to tell him what to do "for his own good." He began to lose his ability to distinguish metaphor or irony from signs of danger. He began to feel that when he himself attempted to convey the irony of his position—that he was a true liberal being acclaimed by the rednecks, as if he were a redneck himself—he found that people were taking him literally. He could not articulate ironically any more, and his wit and sense of humor failed him.

He found himself totally preoccupied with feelings of humiliation. The more deeply he felt reviled, the less he was able to distinguish the source of his social humiliation and shame. He felt during this period that the left side of his head began literally to ache. He felt that a change was occurring inside his head. He became preoccupied with himself as a bad person, a man unwanted and repulsed. The more he felt this way, the more he perceived insult in every remark directed to him. As his feelings of humiliation, rage, and hurt took him over, he was unable voluntarily to attend to his surroundings. He became increasingly forgetful. He would begin tasks and find himself hours later deeply engaged in his pain. While in these fuguelike states he would sometimes emerge for a few minutes to try to connect with the task at hand. The intense significance he experienced at those times affirmed his feeling that people were against him. And indeed, people were beginning to talk about him, to doubt his competency. They would stop in front of him, wave their hands in front of him, call him by name and wonder aloud if this could be the Mr.—— they had come to know and respect as their spokesman in a time of political turmoil. He could only hear derogation in their remarks.

Then he had a dream: the world was coming to an end; he had a special mission. He was to be a second Kennedy. He would save the world from the malicious destroyers. In the aftermath of the dream the patient became convinced of a vast conspiracy to hurt and destroy him, he didn't know why. In those moments when he emerged from his preoccupation with pain and delusion, he looked for significance in every remark addressed to him. He began to plant trick phrases in his speech in order to distinguish those who were from those who were not involved in the conspiracy. He waited to see which persons brought these phrases back to him. Gradually even allusions to the phrases were sufficient proof of conspiracy. He found an everwidening circle of conspirators. He hired a detective to try to find out how the linkages among the conspirators were made and to investigate those persons he believed behind the conspiracy.

At this point he began to notice that his mind was invaded by forced memories of past events. Since these events involved his past and his family, he felt that the conspiracy had come to involve family members. He believed that a gossipy acquaintance of his who knew his cousin must have spread the conspiracy to his family. When another relative did spontaneously talk to him about the difficulty he had heard the patient was in, he was sure that his whole family, or most of it, had become involved in the conspiracy. Then a new revelation was added to the conspiracy theory. He began to think that his father was not really his father. He confronted his parents about this, and his parents admitted that his mother had had an affair with her sister's husband, his uncle. His father had been cuckolded. Then he was sure that his uncle or some other unknown person had fathered him. This added fuel to his family romance fantasy—he did have different parentage. Now his past began to make sense to him in a new way. He began to believe that all of his suffering had come, as indicated in the dream, for the purpose of saving the world from political and social destruction. As the forced recollection from the past continued, he felt that his mind was being taken over by an alien force, and he felt himself rapidly weakening. He feared that he would lose his mind entirely.

Treatment Unable to bear the torment and terror of falling from social grace, acute psychotics develop false, delusional identity structures to compensate for their lost social positions. The false object structures they achieve belong to their attempts to restabilize their identities at any cost, for the pain and intense anxiety of the psychosis are unbearable. These efforts lead to delusions and permanently laughable identity structure that cannot achieve the social restitution they are meant to create.

The man now being discussed came into treatment in this most inimical stage of psychosis. Treatment therefore aimed at halting his delusion formation before it became a fixed part of his personality structure. Treatment began with medication. He was placed on a combination of amitriptyline, 150 mg and trilafon 12 mg (triavil 4-50, iii tabs h.s.) as soon as he could tolerate this dosage. I wanted the elavil to stabilize the norepinephrine and serotonin systems so that the patient could process reality more easily and with less irritation. I wanted the trilafon to cut down on the drive pressure for consummation that continued to goad him to new action. This combination of drugs is often used in delusional illness with a depressive component because the blend produces less movement disorder than other mixtures do.

I took the patient's tolerating this much medication as a sign that he did indeed have a stress-related, depressive component to his illness. In my experience, elavil and other antidepressants exacerbate the psychosis of patients who have a purely paranoid type of schizophrenia. I also believe that in chemically inducing a chronically delusional paranoid patient's attention to reality, antidepressant medication exacerbates the condition. The forced attention to reality exposes the disjunction between delusional character structure and reality, triggering the formation of further delusion formation. The false character does function to provide a shield against the terror and pain of existence in a hostile social world. The nonappearance of any new pathology tended, therefore, to confirm my hypothesis that the paranoia consisted of a stress-related, nondominant type of response.

The fact that the patient tolerated the medication and that he began to sleep better was, for me, a clinical indication that it would be worthwhile to attempt to reverse the psychotic process altogether, both by analyzing it and by providing reconstructive explanation of the psychopathological changes the patient had undergone. Thus, in building the frame of treatment, I emphasized what the patient was already emphasizing: he had held onto his wife and sought my help, two indicators that he did not want to succumb to psychosis. In this way I hoped to promote a vertical split in which the therapist is allied with the healthier identity functions of the patient; and this can be used to build upon, in order gradually to reclaim the identity components and ego functions which had been enlisted in the psychotic process.

To facilitate the patient's allying his identity structure with the therapist's, as well as the therapeutic method of using this alliance to strengthen the function of social judgment that has been so badly impaired in cases of paranoia, one must agree as much as is possible with the patient's account of his or her mistreatment and persecution. As was true in this case, the chances are that the patient was even more misused, abused, and mishandled than he or she reports. There is not a mere nucleus of truth at the center of the paranoid's delusions; there is a whole city of truth. This truth always relates to the acute break with reality that ushers in the psychosis. Subsequent breaks are facilitated by the patient's open delusional structure, more than they are triggered by social maltreatment.

In this case I was able to sympathize deeply with the patient's mistreatment by his essentially paranoid milieu. Seeing his early actions in this episode—and even some of his later ones—as courageous, I was able to tell him that he had exhausted himself by trying day and night to restore his social place, while also laboring to change the false social structure around him. I explained that when it is unsuccessful, such an effort yields the constant turmoil and humiliation he had experienced. This had finally exhausted him and impaired his ability to concentrate. I could also tell him that his family history had left him particularly vulnerable to this kind of humiliation. As he reminded me of portions of his family history that I had repressed or forgotten or failed to understand in the past, I could point out the repetition compulsion with a feeling of real excitement at the discovery. I could point out that he had emerged from a whole childhood of humiliation and hopelessness to an adult life he had never believed possible. He had overcome social isolation once in his life and he could do so again.

The paradigm helped me to understand and therefore communicate to him how his withdrawal into humiliation had distanced him from reality, producing difficulties in the integration of reality. I pointed out that when he found his ordinary social self unrecognized he was unable to organize his life in the world, and that he began to look for significance in events in order to hold onto a reality that was failing him. I was able to add that the discontinuity in events brought about by his wavering attention, and his withdrawal, in combination with his intensified experience of significance, had led to the disturbing changes he experienced in his sense of time and in his ability to locate events, for he had condensed events of two summers and three investigations so badly that he despaired of ever recovering the facts.

This brings us to the fact that cognitive explanations of pathological events are an important part of the therapy of psychosis. When the patient can understand the chain of events that led to altered experience, he can retrace the steps that led to the pathology. The experience of cognitive changes convinces a patient that he is losing his mind, and this experience is a major trigger for delusion formation, for cognitive integrity requires that all experience be explained. Explaining intrapsychic events to the patient reverses the process of developing pathological generalizations to explain the altered experience. Many pathological generalizations about psychosis have become part of the mythology of mental illness, both for the general public and even for the practitioner. A patient may believe, for example, that he has had a nervous breakdown or developed a split personality. Both statements contain some approximation of the truth, but neither leaves the patient with any way of exploring the real events and their impact that led to the psychosis. Supporting the false generalization of identity that society offers to explain psychosis may help to maintain stability in chronic psychosis, but it hinders any recovery process.

Characteristically, the patient began each session by bringing up something he felt I had misunderstood about his experience in the previous session. Inevitably, the ambiguity hinged on my failure to understand some aspect of his experience that revealed real mistreatment. In addition, he wanted the interpretation and reconstruction repeated in order to feel sure about relinquishing a piece of delusion. Each disclaimer he made of my reconstruction added a piece for further reconstruction of his psychotic process and of the events that had fed into it. Finally I would restate the previous session's reconstruction, adducing a further understanding of the delusional content. As this work continued, the weight of alternative reconstruction began to overweigh the whole conspiracy theory. It began to be possible to take on the whole conspiracy and to explain its intrapsychic function.

To be more specific, during his humiliation he was so preoccupied that he was unable to attend to events. He maintained mental order and cognitive order during his preoccupation with disgrace by arriving at simplified generalizations about his pain. These simplified and transformed the experience in a way that he could accept. I pointed out (as one could point out to any psychotic patient) that he had begun to ignore other people's complex qualities, as well as his own, through this process of generalization. He increasingly had come to believe in conspiracy as a way of unifying the results of other people's actions toward him. He saw them as all bad. Believing that he deserved this treatment because his earlier life had conditioned him to feel that he himself was all bad, all unworthy, he tried to protect himself from this painful generalization by developing the alternative delusional belief that he was all good. Thus, he arrived at the delusional generalization that he was being persecuted to save mankind. The reconstruction of these changes must be made, of course, in pieces, and repeated, because it challenges the nature of the secondary delusions that are themselves being used to reconstruct a new delusional identity. In order to accept such interpretations and cognitive reconstructions, the patient must reimmerse himself in the pain and humiliation of the original experience. At that point, in the crucible of the psychotic-making experience the therapeutic alliance plus the patient's interest in reality must be strong enough to hold against the patient's choosing to make the therapist a new conspirator. In the sessions with this patient he would reimmerse himself in the humiliation as soon as the initial work of the session was completed. He would show me the pain and express his panic that he could not emerge intact from the experience.

Once the structure of the psychosis has been understood the patient must naturally begin to want to restore his life. This means reapproaching the social vulnerability in a new way through re-entering the social world. In this patient's case, it became necessary for him to find another job because he and his wife were sinking into debt. As part of the process of looking for another job he applied for an academic librarian's position at a large college. He felt tremendous pressure about the application because he did not know how he was going to account for what had happened to him in his last job. Why had he resigned from a tenured position? It might seem easy enough for him to tell about the various investigations, but talking about the incidents would throw him into a state of pain and humiliation from which he would not be able to extricate himself during the interview. Nor could he, when he anticipated the interview, keep himself from believing—or at least entertaining—the conspiracy ideas. In the thrall of these ideas, he even feared that the interview committee might be in on the conspiracy.

This spreading fear of conspiracy often causes paranoid people to pack up and move elsewhere. When they arrive elsewhere their faulty restitutions lead to further feelings of humiliation and lack of social acceptance, and this leads to the belief that the conspiracy has followed them, caught up with them in the new place. Someone is always bound to have known someone in the previous place, or in some past place, and that, or some connection even more incidental, is enough to persuade the paranoid person that the conspiracy has spread.

Trying to manage such feelings, the patient went off for his interview. During the interview, and particularly when he was accounting for his decision to leave his last job, the stress became such that he felt an intense pain in the left portion of his head, a pain that moved back and forth in the frontal area.

I believe that such symptoms are not rare, that they occur when a patient is using all of his prefrontal mental resources. This produces an overflow of blood and neurotransmitters to the affected prefrontal area. The neural intensity produces the maximum neurotransmitter response, which in turn releases painful exudates that surround the overworking neural areas. In this case, I surmise that the intense planful effort, the feeling of humiliation, and the panic that he could literally be seen through and known for his role in the previous investigations was so great that he could not continue his application for the job. He did realize that the interviewers could see his hands shaking; and he could not know how much more they knew. Thus, the attempt to renew his social and occupational life at a high level precipitated anew acute regression.

The work of therapy, however, was not undone. The patient simply concluded that he could not get such a prestigious job while still suffering from his illness. I did explain his headache in the present terms. Such events and the accompanying physical sequelae often keep chronically psychotic people from breaking through to a higher level of functioning. When too much social participation is required, the whole array of regression-triggering feelings return like monsters from the deep.

Soon after the job interview, the patient reported the first dream of his therapy. He was entering a room that he realized was my office. He was conducted into this session or interview room by a kindly man, but once he passed over the threshold, the whole scene changed. The room became enveloped in a fire that was emanating from his conductor, who had taken on the shape of a huge lizard. Attempting desperately to escape from this fire-breathing man, the patient finally managed to change himself into the commanding figure of John F. Kennedy.

Having the structure of the psychosis brought to life in this way made it possible for us to do further therapeutic work. Clearly, in addition to representing the difficult interview, the room in the dream also represented the patient's fear of the therapy situation. In the therapy he must engage his psychosis which he portrayed as a consuming, monstrous fire of pain that overwhelms him. After this was interpreted, along with the evident wish that the patient had to transform himself into a powerful political figure to strengthen himself and to avoid pain, and after I sympathized with how difficult it must be for him to engage in therapy under the circumstances, the patient said that I should go ahead and be direct with him, that he would be able to take it, and do the necessary work. Although the homosexual masochistic theme is very apparent in his 'stick it to me' reaction, adherence to the more neutral therapeutic alliance is a much stronger part of the message. In the dream he is allowing us both to share the knowledge of the dynamics and structure and his psychosis, because he wants to work it through and to understand it. After this discussion we were able to go farther and to talk about this aspect of his character—his grandiose kind of masochism, his making the best he could out of his persecution—as an aspect of his therapy that we could take up after he was safely able to understand his psychotic episode and to encompass it within the framework of his present experience. After one year he confessed his belief that I was a benevolent conspirator. This had given him the strength to trust me. Relinquishing the psychosis and the psychotic transference left a severe secondary depression in its wake.

CONCLUSION

As therapists, we can interpret, reconstruct, and explain to the patient only as much as we understand about his mental life. I am firmly convinced that the present paradigm does enhance our observational powers, for it links them to a newly perceived set of phenomena within the individual. The understanding of the neurological roots and neuropsychological structure of the human mind cannot fail to give rise to improvements in therapeutic technique. Just as Freud's observations of childhood sexuality in its broad scope gave rise to our ability to see childhood masturbation and the shaping effects of childhood fantasy life more clearly than before, just as Kohut's power of observation of the subjective self that longs for mirroring confirmation throughout life gives rise to an enhanced observation of the vicissitudes of human consciousness, so also does the neurological approach outlined in this book give rise to an enhanced understanding of the organization of consciousness.

I feel that the structural paradigm has changed my approach to the treatment and understanding of my patients. Taking the last patient described as an instance, I recall that earlier in my career, when I treated him in his adolescence and young adulthood, I interpreted his symptoms along the lines of my psychoanalytic understanding. I understood, that is, and so I interpreted, that his narcissistic development had been blocked by his fear of his homosexual wishes. I also understood and interpreted his competitive strivings in terms of the regressive transformation of his wishes because of intense castration anxiety that entered into his relationship with his father, his brothers, and cousins. Thus, because he feared that all of the men in his life were weak in comparison with his mother and the powerful sister, he expressed his sexual competitive wishes in terms of a masochistic identification with his father as the weak or enfeebled, fragile male. His persecution fantasies were the regressive transformation of his competitive wishes when faced with castration threats from the side of his mother, aunt, and his brothers.

In my present view these analytic formulations of his problem must be further explained. I now believe that this patient's narcissistic vulnerability leads beyond mere psychic regression to forced painful experiences of humiliation and terror that include memories of his past that come over him in transforming waves to set a psychotic syndrome into motion. Indeed, the patient told me that long after he stopped seeing me, he would have episodes in his twenties and early thirties when he experienced memories forced into his consciousness. He has not yet elaborated on their content, but in the context of the present paradigm I must take these episodes as evidence of regression that proceeds (further back than mere psychological regression) into the realm of kindling and forced personality change. One might say that a temporal lobe process of partial epilepsy is at work, but the occurrence of these regressive kindlings as the end result of increasing conflict speaks for their psychological origin.

Certainly we must revise our notions of what is psychological and what is neurological in origin. This conclusion relates to our understanding of what a genetic disposition to mental illness means. Certainly we can see that the patient's father has a disposition to paranoia. We might see this disposition to the formation of a paranoid illness as a greater neurological predisposition to kindling, given a certain level of intrapsychic conflict.

Thus, where I once understood the development of significance experience and the alteration of identity based on that significance as evidence for Freud's hypothesis of restitutional symptom formation in psychosis, I now see that process in its neurological terms as well. I see this man's psychotic process as emerging first in the breakdown of upper level identity functions that reduce the inhibiting capacity of the prefrontal cortex, and then secondarily in the consequential emergence of kindling processes that force a new kind of experience of reality into being. This new experience must consequently, and syndromatically, trigger the formation of new psychic structures for encompassing the new experience. Finally, given the biological need to prevent outbreaks of dangerous behavior brought about by the cortical disinhibition and by the formation of new versions of reality, it is necessary for the psychotic individual to form new, restitutional structures of intrapsychic identity based on the altered experience. These new identity structures replace the failed ones and allow for a renewed functionality at the higher levels, albeit a much more socially dysfunctional identity structure. The delusional explanations engendered by the false structures of identity are invariably based on the most gross, but nevertheless inclusive, generalizations of experience. These circumstantial, vague rationalizations, alluding to the whole psychotic process, provide sufficient integration of identity to inhibit the formation of emotionally painful new regressive kindling, at the same time that they provide a "sick person" or "crazy person" identity structure that is understandable by a society that stops expecting the patient to function independently.

A further look at the restitutional or progressive aspect of the psychotic process, provide sufficient integration of identity to inhibit the formation of emotionally painful new regressive kindling, at the romance fantasy as a fantasy of compensation that paranoids and other creative individuals used to re-enforce their grandiosity in a compensatory fashion. The fantasy of having a different set of parents accounts for the feeling of personal greatness encountered in such individuals. Freud also understood that in paranoia the disintegration of identity decomposes earlier identifications formed during the course of life. Thus Freud was in the position to understand, although he never said it, that the family romance identity alteration is a compensatory restitution for the whole breakdown of identifications that occurs in the psychotic process.

Through our understanding that each stage of development

forms a new structure of identity based on kindling process, we may understand that the psychotic restitution with the family romance is a natural, if pathological, form of identity formation in which extreme identity generalization forms an inclusive new identity structure in which all previous organizing fantasies are included and all previous developmental kindling processes are inhibited. Thus, if we go back to the eighteen month primal scene organization of identity, we can see that the primal scene denial, positing that the parents never did join together in exclusive encounters, is revived in the family romance. Thus it is some other set of parents that coupled, not the child's own. Therefore there is nothing to be afraid of in reality and no need to give up primal symbiotic fantasies of union. In the family romance fantasy the child is forever immune from the disturbing realization that reality has its own set of rules and regulations that exclude the child's infantile wishes. The parents and society as an extension of the parents never did have that exclusive encounter that left the child out in the psychic cold.

In paranoia, the inclusion of the family in the delusional structure of conspiracy, and the compensation for that through the formation of a grandiose restitutional identity sometimes incorporating family romance elements, is usually a sign of chronicity and irreversible psychotic process. In the light of such a massive restructuring of the organizational aspects of identity old memories are reorganized to conform to the new structure of identity. One can easily imagine the patient's concluding that his aunt was all along behind the social conspiracy that robbed him of his capacity to be a great man.

One further expansion of our context due to the neurological approach relates to the developmental stage that shapes the conflict giving rise to the pathological syndrome. The present patient approaching his late thirties has the developmentally mandated task of coming to grips with the reduction in the availability of neurotransmitter resources. We understand that the individual must at this stage begin to redefine the self and personal identity in terms of social structure at this juncture of life. This is the period of life when paranoid patients or those with paranoid tendencies often engage in deep conflict with society because they can hardly reshape their personality in accord with social norms when they feel that they have been at odds all along with social structure. The acceptance of decreased permanent self-esteem due to a feeling of becoming a permanent misfit in some way in society is a contributor to the formation of paranoia at the time of entrance to the mature years of life.

REFERENCES

Abrams, R., & Taylor, M.A. Differential EEG pattern in affective disorder and schizophrenia. *Archives of General Psychology,* November 1979, *36.*

Adamac, R.E., & Adamac, E.S. Limbic kindling and animal behavior and implications of human psychopathology associated with complex partial seizures. *Biological Psychiatry,* 1983, *18,* No. 2.

Ainsworth, M.D.. *Patterns of attachment.* Hillsdale, NJ: Lawrence Ehrlbaum, 1978.

Akiskal, H.S. Dysthymic disorders: Psychopathology of proposed chronic depressive subtypes. *American Journal of Psychiatry,* January 1983, *140,* 1.

Ariel, R. Regional cerebral blood flow in schizophrenia, *Archives of General Psychiatry,* March 1983, *40.*

Averill, J.R. A selective review of cognitive behavioral factors involved in the regulation of stress. In: R.A. Depue (Ed.) *The psychobiology of the depressive disorders.* New York: Academic Press, 1979.

Bechtereva, N.P. *The neuropsychological aspects of human mental activity.* New York: Oxford University Press, 1978.

Berg, W.K. & Berg, K.M. Psychophysiological development in infancy. In: J.D. Osofsky (Ed.) *Handbook of infant development.* New York: Wiley, 1979.

Bondareff, W. Neurobiology of aging. In: J. Birren & B. Sloan (Eds.) *Handbook of mental health and aging*. Englewood Cliffs, NJ: Prentice Hall, 1980.

Brazelton, T.B. Specific neonatal measures: The Brazelton neonatal assessment scale. In: J.D. Osofsky (Ed.) *Handbook of infant development*. New York: Wiley, 1979.

Brazelton, T.B. The origins of reciprocity. In: S.I. Harrison & J.F. McDermott Jr. (Eds.) *New directions in childhood psychopathology*. New York: International Universities Press, 1980.

Brown, G.L. Aggression, suicide and serotonin: Relationships to CSF amine metabolites. *American Journal of Psychiatry*. June 1982, *139*, 6.

Brown, G.W. The social etiology of depression. In: London Studies in *The psychobiology of the depressive disorders*. R.A. Depue (Ed.). New York: Academic Press, 1979.

Buchsbaum, M. Neurophysiological reactivity, stimulus intensity modulation, and the depressive disorders. In: R.A. Depue (Ed.) *The psychobiology of the depressive disorders*. New York: Academic Press, 1979.

Bunney, E. Jr., Goodwin, F.K., Murphy, D.L., House, K.M. & Gordon, E.K. The "switch process." In: Manic-depressive illness. *Archives of General Psychiatry*. September 1972, *27*.

Carey, S. & Diamond, R. Maturational determinants of the developmental course of face encoding. In: (Eds.) *Biological studies of mental process*. Cambridge, MA: MIT Press, 1980.

Carlson, P. Occurrence, distribution and physiological role of catecholamines in the nervous system. *Pharmaceutical Review*. 1959, *11*, 300–304.

Carroll, B.J. A specific laboratory test for the diagnosis of melancholia. *Archives of General Psychiatry*. January 1981, *38*.

Charney, D.S. Abrupt discontinuation of tricyclic antidepressant drugs: Evidence for noradrenergic hyperactivity. *British Journal of Psychiatry*. 1982, *141*, 377–386.

Chouinard, G., Young, S.N. & Annabe, L. Antimanic effects of clonazepam, *Biological Psychiatry*. 1983, *18*, 4.

Cohler, B.J. Adult developmental psychology and reconstruction in psychoanalysis. In: S. Greenspan & G. Pollock (Eds.) *The course of life*. Washington, D.C.: NIMH, 1980.

Cowdry, R.W. Thyroid abnormalities associated with rapid cycling bipolar illness. *Archives of General Psychiatry*. April 1983, *40*, 414–426.

Crease, I. The classification of dopamine receptors: The relationship to radioligand binding. In: Maxwell Cowan (Ed.) *Annual review of neuroscience*. Palo Alto, CA: 1983.

Diagnostic and statistical manual III (3rd ed.). Washington, D.C.: American Psychiatric Association, 1980.

Doering, C.H. The endocrine system. In: O.G. Brim Jr. & J. Kagan (Eds.)

Constancy and change in human development. Cambridge, MA: Harvard University Press, 1980.

Eccles, J.C. How the self acts on the brain. *Psychoneuroendocrinology*, 1982, *3*, no. 4, 271–283.

Emde, R.N., Gaensbauer, T.J. & Harman, R.J. Emotional expression in infancy. *Psychological Issues*, New York: International Universities Press, 1976. *1*, Monograph 37.

Everitt, A.V. & Huang, C.Y. Aging. In: J. Birren & B. Sloan (Eds.) *Handbook of mental health and aging*. Englewood Cliffs, NJ: Prentice Hall, 1980.

Flor-Henry, P. On certain aspects of the localization of the cerebral systems regulating and determining emotion. *Biological Psychiatry*, 1979, *4*, 677–694.

Freud, S. *Project for a scientific psychology*, J. Strachey, Ed. and trans. Standard Edition, (Vol. I).

Freud, S. (1900) *The interpretation of dreams*, J. Strachey, Ed. and trans. Standard Edition, (Vols. IV & V).

Freud, S. (1905) *Jokes and their relation to the unconscious*, J. Strachey, Ed. and trans. Standard Edition, (Vol. VIII).

Freud, S. (1909) *Family romance*, J. Strachey, Ed. and trans. Standard Edition, (Vol. IX).

Freud, S. (1911) *Psychoanalytical notes on an autobiographical account of a case of paranoia*, J. Strachey, Ed. and trans. Standard Edition, (Vol. XII).

Freud, S. (1917) *Mourning and Melancholia*, J. Strachey, Ed. and trans. Standard Edition, (Vol. XIV).

Freud, S. (1918) *From the history of an infantile neurosis*, J. Strachey, Ed. and trans. Standard Edition, (Vol. XVII).

Freud, S. (1919) *The uncanny*, J. Strachey, Ed. and trans. Standard Edition, (Vol. XVII).

Freud, S. (1923) *The Ego and the Id*, J. Strachey, Ed. and trans. Standard Edition, (Vol. XIX).

Freud, S. (1925) *Negation*, J. Strachey, Ed. and trans. Standard Edition, (Vol. XIX).

Gershon, E.S., & Buchsbaum, M. A genetic study of average evoked response augmentation in affective disorders. In: A.J. Shagass, E.S. Gershon & A. Friedhof (Eds.) *Psychopathology and brain dysfunction*, New York: Raven Press, 1977.

Giannitrappini, D., & Kayton, L. Schizophrenia and EEG spectral analysis. *EEG and Clinical Medicine*. 1974, *36*, 377–386.

Gloor, P., Olivier, A., Quesney, L.F., Andermann, F. & Horwicz, S. The role of the limbic system in experiential phenomena of temporal lobe epilepsy. *Annual of Neurology*. 1982, *12*, 129–144.

Glowinski, J. Some properties of the ascending dopaminergic pathways: Interactions of the nigrostriatal dopaminergic system with other neuronal pathways. In: F. Schmidt (Ed.) *The Neurosciences*. Cambridge, MA: MIT Press, 1979.

Goddard, G.V., McIntyre, D.C., & Leech, C.K. A permanent change in brain function resulting from daily electrical stimulation. *Experimental Neurology*. *25*, 1969.

Gold, M.S., Redmond, D.G. & Heber, H.D. Noradrenergic hyperactivity in opiate withdrawal supported by clonidine reversal of opiate withdrawal. *American Journal of Psychiatry*. January 1979, *136*, 1.

Goldberg, A. & Gedo, J. *Models of the mind*, Chicago, IL: University of Chicago Press, 1973.

Gruber, H.E., & Voneche, J. *The essential Piaget*. New York: Basic Books, 1977.

Grunes, J. Reminiscences, regression and empathy. In: *The course of life*. (Vol. III). Washington, D.C.: NIMH, 1980.

Gunderson, J.G., & Englund, K.J. The families of borderlines. *Archives of General Psychiatry*. 1980, *37*, 27–33.

Gunderson, J.G., Siever, L.J., & Spaulding, E. The search for a schizotype. *Archives of General Psychiatry*. January 1983, *40*.

Harris, J., & Harris, J. The one-eyed-doctor: Sigmund Freud, New York: Jason Aronson, 1984.

Hartmann, H. *Essays on ego psychology*. New York: International Universities Press, 1964.

Henry, J. & Ely, D.G. Etiological and physiological theories. In: I. Kutash (Ed.) *Handbook on stress and anxiety*. San Francisco, CA: Jossey Bass, 1980.

Ingvar, D.H. Distribution of cerebral activity in chronic schizophrenia, *The Lancet*, December 21, 1974.

Ingvar, D.H., Abnormal distribution of cerebral activity in chronic schizophrenia. In C.F. Baxter & T. Melnechuk (Eds.), *Perspectives in schizophrenia research*. New York: Raven Press, 1980.

Inhelder, B. & Piaget, J. *The growth of logical thinking*. New York: Basic Books, 1958.

Insel, T.R., Obsessive-compulsive disorder: A double blind trial of clomipramine and clorgyline, *Archives of General Psychiatry*, 1983, *40*, 605–606.

Itil, T.M., Nootropics: Status and prospects, *Biological Psychiatry*, 1983, *18*, 5.

Jacobson, E. *The self and the object world*, New York: International Universities Press, 1964.

Janis, I. & Mann, L. *Decision making*, New York: The Free Press, Macmillan, 1977.

Kato, G. & Ban, T. Central nervous system receptors in neuropsychiatric disorders, *Prog. Neuro. Psychopharmacology and Biological Psychiatry*, 1982, *6*(3) 207–222.

Keating, D.P., *Thinking processes in adolescence.* In: J. Adelson (Ed.) *Handbook of adolescent psychology.* New York: Wiley, 1980.

Kendler, K.S., Gruenberg, A.M. & Strauss, J.S., An independent analysis of the Copenhagen sample of the Danish adoption study of schizophrenia: II. The relationship between schizotypal personality disorder and schizophrenia. *Archives of General Psychiatry,* September 1989, *38.*

Kestenberg, J. (a) Eleven, twelve and thirteen: Years of transition from the barrenness of childhood to the fertility of adolescence. In: S. Greenspan, & G. Pollock (Eds.) *The course of life* (Vol II). Washington, D.C.: NIMH, 1980.

Kestenberg, J. (b) Maternity and paternity in the developmental context. In: *The Psychiatric Clinics of North America,* Sexuality. (Vol III) April 1980, 76–77.

Kety, S.S., Rosenthal, D., Wender, P.H. Mental illness in the biological and adoptive families of adopted individuals who have become schizophrenic: A preliminary report based on psychiatric interviews. In R. Fieve, R. Rosenthal, H. Brill (Eds.) *Genetic Research in Psychiatry.* Baltimore, MD: Johns Hopkins University Press, 1975, 147–165.

Kihlstrom E. Summation. In: R.A. Depue (Ed.) *The psychobiology of depression.* New York: Academic Press, 1979.

Kohut, H. *The restoration of the self.* New York: International Universities Press, 1977.

Kornhuber, H. Chemistry, physiology, and neuropsychology of schizophrenia, *Archiv fur Psychiatrie und Nervenkrankheiten* 1983, *6*(233) 415–422.

Kuhn, T.S. *The structure of scientific revolutions,* Chicago: University of Chicago Press, 1962.

Laing, R.D. *The divided self.* London: Tavistock Publications, 1959.

Lauffer, M. The central masturbation fantasy, the final sexual organization and adolescence. In: *The psychoanalytic study of the child.* 1976, *31.*

Leonard, L.B. *Meaning in child language.* New York: Grune & Stratton, 1976.

Luchins, D., Weinberger, D.R., & Wyatt R. Schizophrenia and cerebral asymmetry detected by computed tomography. *American Journal of Psychiatry,* June 1982, *139,* 6.

Luria, A.R., *Higher cortical functions in man.* New York: Basic Books, 1980.

Maas, J.W., Hattox, S.E., Greve, M., & Landis, D.S. MHPG production by human brain in vivo. *Science,* 7 September 1979, *205,* 1025–1027.

MacLean, P. Phylogenesis. In: P. Knapp (Ed.) *The expression of emotions in man.* New York: International Universities Press, 1963.

Mahler, M., Pine, F. & Bergman, A. *The psychological birth of the human infant.* New York: Basic Books, 1975.

Martin, J.B., Reichlin, S. & Brown, G. *Clinical neuroendocrinology*. Philadelphia, PA: F.A. David Co., 1977.

McGeer, P.L., Eccles, J.C. & McGeer, E.G. *Molecular biology of the mammalian brain*. New York: Plenum Press, 1978.

Meltzer, H.Y. Dopamine autoreceptor stimulation: Clinical significance. *Pharmacology, Biochemistry and Behavior,* 1982, *17*, Supplement.

Meltzer, H.Y., Brinda, U., Robertson, A., Tricou, B.J., Lowy, M. & Perline, R. Effect of 5-Hydroxytryptophan on serum cortisol levels in major affective disorders. *Archives of General Psychiatry*, April 1984, *41*.

Michalewski, H.J. Use of the EEG and evoked potentials in the investigation of age-related clinical disorders. In: B. Sloan & J.E. Birren (Eds.) *Handbook of mental health and aging*. Englewood Cliffs, NJ: Prentice-Hall, 1980.

Mountcastle, V. An organizing principle for cerebral function: The unit module and distributed system. In F. O. Schmidt (Ed.) *The Neurosciences*. Cambridge, MA: MIT Press, 1979.

Osofsky, J.D. & Connors, K. Mother-infant interaction: An integrative view of a complex system, In: J.D. Osofsky (Ed.) *Handbook of infant development*. New York: Wiley, 1979.

Paykel, E.S. Recent life events in the development of the depressive disorders. In: R.A. Depue (Ed.) *The psychology of the depressive disorders*. New York: Academic Press, 1979.

Piaget, J. *The construction of reality in the child,* New York: Basic Books, 1954.

Pollock, G.H. Aging or ages. In: S. Greenspan & G. Pollock (Eds.), *The course of life. (Vol. III),* Washington, D.C.: NIMH, 1980.

Pope, H.G., Lipinski, J.F., Cohen, B. & Axelrod, D.T. Schizoaffective disorder: An invalid diagnosis? A comparison of schizoaffective disorders, schizophrenia, and affective disorders. *American Journal of Psychiatry*, 1980, *137*, 921–27.

Pope, H.G. & Hudson, J.I. Bulemia treated with imipramine: A placebo controlled double blind study. *American Journal of Psychiatry,* May 1983(a), *140*, no. 5.

Pope, H.G. The validity of DSM III Borderline Personality Disorder. *Archives of General Psychiatry,* January 1983(b), *40*.

Post, R.M., Fink, E., Carpenter, W.T. & Goodwin, F.K. Cerebrospinal fluid amine metabolites in acute schizophrenia. *Archives of General Psychiatry*, August 1975, *32*.

Pribram, K.H. Emotions. In: S.B. Filskov & T.J. Boll (Eds.) *Handbook of clinical neuropsychology*. New York: Wiley, 1981.

Remillard, G.M., Andermann, F., Franco Tosta, G., Gloor, P., Aubé, M., Martin, J.B., Feindel, W., Guberman, A., & Simpson, C. Sexual ictal manifestations predominate in women with temporal lobe epilepsy: A finding suggesting sexual dimorphism in the human brain, *Neurology*, March 1983, *33*, 323–30.

Renner, J. & Birren J. Stress: Physiological and psychological mechanisms. In: J. Birren & B. Sloan (Eds.) *Handbook of mental health and aging*. Englewood Cliffs, NJ: Prentice-Hall, 1980.

Richards, J.G. & Mohler, H. Benzodiazepine receptors. *Neuropharmacology*, 1984, *23*, no. 2B 233–42.

Rose, R.M. Endocrine responses and stressful psychological events. In: *The Psychiatric Clinics of North America*, 1980, *3*, 2.

Rosen, I. Basic neurophysiological mechanisms in epilepsy. *Acta Psychiatrica Scandinavica*, 1984, Supplementum no. 313, *69*.

Rubinow, D.R. CSF somatostatin in affective illness. *Archives of General Psychiatry*, April 1983, *40*.

Saric, R.H. Neurotransmitters in anxiety. *Archives of General Psychiatry*, 1982, *39*.

Satinoff, E. Are there similarities between thermoregulation and sexual behavior? In: D.W. Pfaff (Ed.) *The physiological mechanisms of motivation*. New York: Springer Verlag, 1982.

Scheflen, A., Brsulak, P.J., Schatzberg, A.F., Gudeman, J.E., Cole, J.O., Rohde, W.A. & La Brie, R.A. Toward a biochemical classification of depressive disorders. *Archives of General Psychology*, December 1978, *35*.

Schonfeld, D. Learning, memory and aging. In: J. Birren & B. Sloan (Eds.) *Handbook of mental health and aging*. Englewood Cliffs, NJ: Prentice-Hall, 1980.

Selye, H. The stress concept today. In: R.A. Depue (Ed.) *The psychobiology of the depressive disorders*. New York: Academic Press, 1979.

Shagass, C., Roemer, R.A., Straumanis, J.J. & Amadeo, M. Evoked potential correlates of psychosis. *Biological Psychiatry*, 1978, *13*, no. 2.

Shimkunas, A. Hemispheric asymmetry and schizophrenic thought disorder. In: S. Schwartz (Ed.) *Language and cognition in schizophrenia*. Hillsdale, NJ: Lawrence Ehrlbaum, Pub., 1978.

Snyder, S.H. Opiate receptors in the brain. *New England Journal of Medicine*, 1977, *296*, 266–271.

Snyder, S.H. Drug and neurotransmitter receptors in the brain. *Science*, 1984, *224*, 22–24.

Soloff, P.H. & Millevard, J.W. Psychiatric disorders in the families of borderline patients. *Archives of General Psychiatry*, January 1983, *40*, 37–43.

Soto, M. Longlasting hypersensitivity to metamphetamine following amygdaloid kindling in cats: The relationship between limbic epilepsy and the psychotic state, *Biological Psychiatry* 1983, *18*, no. 5.

Sperry, R.W. Hemisphere deconnection and unity in conscious awareness. *American Psychology*, 1968, *23*, 723–733.

Sperry, R.W., Gazzaniga, M.S. & Bogen, J.E. Interhemispheric relationships: The neocortical commissures: syndrome of hemispheric disconnection. In: G.W. Bruyn (Ed.) *Handbook of clinical neurology*. Amsterdam: North Holland Publishing, 1969, 273–290.

Spitzer, R.L., Endicott, J. & Gibbon, M. Crossing the border into borderline personality and borderline schizophrenia, *Archives of General Psychiatry*, January 1979, *36*.

Sroule, A. Socioemotional development. In: J.D. Osofsky (Ed.) *Handbook of infant development*. New York: Wiley, 1979.

Stevens, J.R., Bigelow, L., Denney, D., Lipkin, J., Livermore, A.H., Raascher, F. & Wyatt, R.J. Telemetered EEG-EOG during psychotic behavior of schizophrenia. *Archives of General Psychiatry*, March 1979, *36*.

Stevens, J.R. Stereotypy, spikes, schizophrenia, and seizures, *Biological Psychiatry*, 1982, *18*, no. 4, 407–410.

Strub, R.L. & Black, W.F. *Organic brain syndromes*. Philadelphia, PA: F.A. Davis Co., 1981.

Sulser, F. Regulation and function of noradrenaline receptor systems in brain. *Neuropharmaçology*, Great Britain, 1984, *23*, no. 2B 255–61.

Swanson, L.W. & Sawchuk, L. Hypothalamic integration. In: M. Cowan (Ed.) *Annual review of neuroscience*. Palo Alto, CA: 1983.

Sweeney, D.R. & Maas, J.W. Stress and noradrenergic function in depression In: R.A. Depue (Ed.) *The psychobiology of the depressive disorders*. New York: Academic Press, 1979.

Trimble, M.R. Interictal psychoses of epilepsy. *Acta Psychiatrica Scandinavicca*, 1984, Supplementum #313, *69*.

Vygotsky, L.S. *Thought and language*. Cambridge, MA: MIT Press, 1962.

Vygotsky, L.S., *Mind in society*. Cambridge, MA: Harvard University, 1978.

Wagner, D.R., Weitzmann, E. Neuroendocrine secretion and biological rhythm in man. In: *The psychiatric clinics of North America*. August 1980, *13*:2.

Weiss, J.M. Coping behavior and stress induced behavioral depression: Studies of the role of brain catecholamines. In: R.A. Depue (Ed.) *The psychobiology of the depressive disorders*. New York: Academic Press, 1979.

Welford, A.T. Sensory, perceptual, and motor processes in older adults. In: J. Birren & B. Sloan (Eds.) *Handbook of mental health and aging*. Englewood Cliffs, NJ: Prentice-Hall, 1980.

Winnicott, D.W. *The maturational processes and the facilitating environment*. New York: International Universities Press, 1965.

Wolpert, E.A. Comments. In: S. Greenspan & G. Pollock (Eds.) *The course of life*. Washington, D.C: NIMH, 1980.

INDEX